THE SMART GUIDE TO

Practical Math

BY JIM STEIN

SECOND EDITION

The Smart Guide To Practical Math - Second Edition

Published by

Smart Guide Publications, Inc.
2517 Deer Chase Drive
Norman, OK 73071
www.smartguidepublications.com

For information, address: Smart Guide Publications, Inc. 2517 Deer Creek Drive, Norman, OK 73071

SMART GUIDE and Design are registered trademarks licensed to Smart Guide Publications, Inc.

International Standard Book Number: 978-1-937636-63-0

Library of Congress Catalog Card Number: 2012933750
11 12 13 14 15   10 9 8 7 6 5 4 3 2 1

Printed in the United States of America

Cover design: Lorna Llewellyn
Copy Editor: Ruth Strother
Back cover design: Joel Friedlander, Eric Gelb, Deon Seifert
Back cover copy: Eric Gelb, Deon Seifert
Illustrations: James Balkovek
Production: Zoë Lonergan
Indexer: Cory Emberson
V.P./Business Manager: Cathy Barker

ACKNOWLEDGMENTS

There are several people who made substantial contributions to this book. First is my agent, Jodie Rhodes, now retired, who brought Smart Guide Publications to my attention and talked me into writing one when I was dubious. Had she not done so, there would be no *Smart Guide to Practical Math*—or at least no book with that title and with yours truly as the author. Second is retired Prof. Les Axelrod of Illinois Institute of Technology. Les responded to all my questions about electricity, an area in which I am remarkably deficient—and if there are mistakes in the electricity chapter, they are mine, not his. Third is Sherry Skipper-Spurgeon, who made numerous down-to-earth suggestions that helped bring the *Practical* into *Practical Math*, especially with regard to everyday problems. Last, but by no means least, is my wife, Linda, who put up with an inordinate amount of obsessive-compulsive behavior on my part while I was writing this book—and helped me by dragging me away from the computer when necessary. She helped me realize that all work and no play makes yours truly—or anyone else—a very dull boy indeed

TABLE OF CONTENTS

INTRODUCTION

This is a very different math book. Its purpose is to enable you to answer the questions that have quantitative answers, such as, "How many pounds of hamburger should I use to make a meat loaf that serves six?" or "At what age should I take Social Security?"

The title of this book is *The Smart Guide to Practical Math*. One definition of the word 'practical' is 'useful and sensible'. Of course, that's relative—what is useful for a plumber may not be so useful for a housewife. As a result, although you will find some material in the book useful and sensible, there will be some material that you would never want or never need. Although you would never use that material, there are other people who almost certainly will.

Most math books aim to teach you math. Although this is highly praiseworthy, especially if you are a student who is considering a career that uses math extensively, such books are really not very useful for the vast majority of people. Most people do not wish to do the analysis necessary to find out how many pounds of hamburger to use to make meat loaf, or to discover at what age they should take Social Security. They want to be able to plug a few relevant numbers into a calculator (or a spreadsheet) and get the answer. That's what this book will enable you to do.

Practical math is the math you need to answer the quantitative questions that are important to you. The answers to those questions are in this book.

PART ONE

The Basics

CHAPTER 1

What Is "Practical Math"?

In This Chapter

➤ How this book will help you

➤ What practical math used to be and what it is now

➤ Standard calculators and percentages

Back in the 1950s, Chuck Berry recorded "School Days," a song about the life of a typical high school student. It included the lines, "American history, practical math, you studyin' hard and hopin' to pass." Practical math back then was a bedrock course in American high schools. You can still find courses in practical math today, but mostly in vocational or trade schools.

What happened to practical math? The one-word answer is *Sputnik*. The threat from advanced technology developed by the Soviet Union created a fear of a missile gap, and the powers that be decided that a concerted effort should be made to push engineering, the sciences, and mathematics. Engineering and the sciences use advanced mathematics, and algebra is the language of advanced mathematics. In an attempt to respond to the technological challenge of Sputnik, practical mathematics got phased out, and algebra got phased in.

Who Needs to Do the Math?

Math teachers would love for you to be able to do the math—to be able to solve the problem by using math to derive the formula that produces the answer. For most of us, though, it's really not necessary. Math answers a broad variety of questions—but most of those questions have been asked over and over again, and the answers have already been determined. Doing the math is necessary for engineers and scientists to solve those questions for which the answers haven't been determined, but for almost everyone who is neither a scientist nor an engineer, doing the math isn't necessary.

Using the Math Is What's Important

The answers to the majority of the questions that require math are already known. So doing the math—solving the problem the way the math teachers want—is reinventing the wheel, big time.

What is also important is to make math practical. That motivates much of what will be presented in this book. For instance, it is clear that anything involving money directly is practical—all financial formulas are therefore practical. But what about such staples as determining the area of a rectangle? That's a pure geometry question, and generally nobody cares about the answer to a geometry problem. However, suppose we were to look at a rectangular field and ask the following questions:

➤ How long does it take to mow it?

➤ How much does it cost to seed it?

Suddenly, geometry becomes practical!

There is also another question that will sometimes be applicable; given certain constraints, what is the best way of doing things? *Best* can mean any one of a number of things such as cheapest or fastest, depending on the problem. This is bottom-line practical math.

I frequently teach a course in business calculus, a course that is almost guaranteed to elicit anguished wails of, "When am I ever going to use this stuff?" I'm only going to tell one anecdote in this book, and here it is:

Sometime in the early 1990s I was teaching business calculus, and there was a knock on my door during my office hours. A man in his early thirties opened the door and told me that he had a story I might enjoy hearing. He told me his name and that he had to drop business calculus. I looked up his grades in my grade book. He was a borderline A student, so he obviously wasn't dropping because of poor grades.

He told me that he owned a small contracting business, and the City of Los Angeles had just put out an RFP (request for proposal) on a contract to wire a building. He noticed that the problem was similar to a well-known optimization problem we had recently discussed in class (you'll see it in Chapter 20), and so he analyzed the wiring for the building using the techniques of calculus, demonstrating that there was no way anyone could do it more economically. He then told me that he had gotten the job (for several million dollars) and was so busy that he simply had no time to continue with the calculus class. I told him that calculus had more than fulfilled its purpose as far as he was concerned.

So when are you going to use the practical math in this book? Hopefully, soon—and often. It's probably too much to ask that you, too, will land a multimillion-dollar contract based on what you learn in this book—but you never know!

Please Tell Me That I Don't Need Algebra

You don't need algebra. If you're a high school graduate, almost certainly the last time that you needed algebra was when you heaved a sigh of relief and put down your pencil on either the SAT or your state's high school exit exam. Once these onerous tasks have been completed, most people never use algebra again—and they don't need to.

But that doesn't mean they don't need math. Math impacts our lives every day, especially when we do anything that requires money.

Practical Math Problem

A Practical Math Question for Everyone

At what age should I take Social Security?

This practical math question impacts every one of us. If you delay taking Social Security, you will get larger checks. One criterion may be to extract the most money from Social Security, but that depends on how long you are going to live. If you have to choose between two Social Security options, it is important to know to how long you need to live in order to get more money from taking Social Security later rather than sooner.

How Much Math Do I Need to Read this Book?

The amount of math you need to know to get the most out of this book varies. Many people will use this book simply for the day-in, day-out business of everyday life: buying and selling things, paying one's taxes, monitoring one's health, maybe doing some home improvements or tending the garden. You don't need very much math to answer all those questions.

On the other hand, many people have a need for technical answers to questions that confront them in the course of their job. They may work in a shop, they may need to know things about electrical circuits or the power needed for an engine to do a certain job. These people will need more math, but they'll probably have to do less in the way of reviewing material because they probably are fairly comfortable with math.

The Good News

You need to know only enough math to be able to read a mathematical formula. So even if you'll need to review some mathematics, what you'll need to review will be at a very elementary level.

Formulas express truths about the relationships between quantities, which is what mathematics is all about. One of the most important questions of everyday commerce is, "What is the total cost of a number of items, all of which have the same price?" We could answer this question using a combination of English and mathematics as follows:

total cost = number of items x price per item.

Mathematics favors using abbreviations in formulas; it saves a lot of space.

Practical Math Problem

Buying a Number of Items at the Same Price

What is the total cost of 3 hamburgers, each of which costs $1.50?

➤ N = number of items

➤ P = price per item

➤ T = total cost

$$T = N \times P$$

The total cost is $4.50.

OK, not rocket science. But the importance of this formula is that it answers all questions regarding the total price of any number of items at any price as long as each item is the same price. This is one of the reasons formulas are so valuable: each formula answers all the possible versions of a question, and all you have to do is plug the numbers into a calculator.

Alternative Forms for Formulas

A formula such as $T = N \times P$ is not just a rule that enables you to compute the quantity on the left side of the equals sign by plugging numbers into the expression on the right side; it tells you how the quantities in the formula (in this case, T, N, and P) are related. Sometimes a little algebra will enable you to solve the equation for the other quantities; sometimes you need a lot of algebra. Or, because in this book—as in life—you don't need algebra, the work can be done for you in an alternative form.

Alternative Forms

$$N = T/P$$

(just to refresh your memory, the / means to divide T by P)

This formula could be used to answer a question such as, "How many hamburgers were purchased if $4.50 was spent on hamburgers that cost $1.50 each?" The answer to this question is that 3 hamburgers were purchased.

$$P = T/N$$

This formula could be used to answer a question such as, "What is the price of a hamburger if 3 of them were purchased for $4.50?" The answer to this question is that the price of a hamburger is $1.50.

Most of the practical math in this book will appear in sidebars such as the one in the "Buying a Number of Items at the Same Price" problem. The answer to the example will appear after the formula that is used to compute it. This will enable you to get practice using the formula and see if you have done the work correctly.

How to Use This Book

The purpose of this book is not to give you a course in practical mathematics. This isn't Chuck Berry's "School Days." The purpose of this book is to give you answers to many of the question that can be resolved through mathematics.

After you get past the first few chapters, which will make sure that you can read the answers and can calculate what you need to know from the relevant formula, the rest of the book is divided into chapters that answer questions from different areas of your life. Since a book is incapable of knowing who purchased it (although possibly that's a concept in development), this book contains the answers to a large number of questions. Some of these questions will be ones you want answered; some won't. The questions are grouped by chapter according to topic. Of course, some questions might appear to belong to any one of a number of topics, so you'll have to look through the chapters.

How to Use the Index

At the end of most books is an index that tells you on what page or pages a certain topic or person appears. The last chapter of this book is a chapter-by-chapter list of the practical math questions that constitute this book, along with the questions in each chapter. Although the In This Chapter section at the start of each chapter also contains a chapter summary, going to the last chapter can save you time and irritation as it will eliminate some of the necessity of flipping through the book.

Do I Need More Than Knowing the Right Formula?

Beyond knowing the right formula, you need to know what everyone under the age of twenty and over the age of five knows: how to plug numbers into a calculator to get the right answer. The right formula is like good advice—it's only valuable if correctly used.

There's another way to calculate answers that's extremely useful, in many ways even more useful than a calculator. A spreadsheet, such as Excel, is an invaluable tool for anyone who finds themselves faced with a situation that requires using a relatively complicated formula over and over again. Bill Gates won't appreciate hearing this, but there are free software packages available for much of the software that Microsoft sells—and you can find them just by browsing the Internet. In order not to antagonize him, however, there will be a section later on that will teach you the basics of Excel.

Can I Skip the Rest of This Chapter?

The rest of this chapter discusses standard and scientific calculators, where you can find them (other than in a store), and the use of the important percent (%) button on a standard calculator. If you know how to use the percent button—or, alternatively, you know how to answer percentage questions without using the percent button—you won't miss anything by going to the next chapter.

Two Types of Calculators

There are basically two types of calculators: standard and scientific. Standard calculators are dirt cheap; you can probably pick one up for less than the price of a latte at Starbucks. Standard calculators are very simple; they enable you to do the four basic arithmetic operations previously discussed, they may have a simple memory capability that will store a number, and they usually have a percent button. A standard calculator will enable you to use a large number of the formulas in this book. A drawback, though, is that some of the formulas are rather lengthy, and you may have to keep track of an intermediate result by writing it down.

The other type of calculator is the scientific calculator. Chances are that you can find a great scientific calculator at an electronics supply store for somewhere in the neighborhood of the price of a latte and a biscotti at Starbucks. I just went to Google shopping and found several. Scientific calculators will enable you to handle any formula that appears in this book.

Getting the Most Out of Your Calculator

Your calculator is a remarkable tool. It is safe to say that we'd probably be decades more advanced in science and technology had calculators first been available in the 1870s rather than in the 1970s. Most people use calculators for only the most elementary operations, such as balancing their checkbook. If you have a scientific calculator, using it just to balance your checkbook is like owning a Ferrari and using it only to drive to the grocery store.

If you are able to use this book, you know how to use a standard calculator with the possible exception of the percent button, and this will be explained shortly. If you own a scientific calculator, it either comes with an instruction booklet or, more likely, the calculator manufacturer has a website that has an instruction booklet (a good way to save on printing costs). It's worth spending a couple of hours running through the basics; you've got a Ferrari in your pocket and you want to know how to nudge it out of first gear.

As in the "Buying a Number of Items at the Same Price" problem, every formula in this book comes with a solved problem. If you get a different answer, you either hit a wrong button accidentally or didn't use the calculator correctly.

Using the Percent (%) Button on a Standard Calculator

Many of the formulas in this book involve percentages. Percentages are an important and useful idea, but they suffer from the liability that most people misunderstand them. That misunderstanding may not start with the percent button on their standard calculator, but it's a symptom of that confusion. Many people do not know how it works.

The percent button on a standard calculator will enable you to do the following important calculations:

➤ Take a percentage of a number

➤ Add a percentage to a number, such as when you tip 15% of the cost of a meal

Subtract a percentage from a number, such as when you buy an item and get a 3% discount for cash

Taking a Percentage of a Number

If you are buying a house and want to take out a loan, an important number is the slightly misnamed loan-to-value ratio. It's slightly misnamed because it's actually a percentage. When a bank says that its loan-to-value ratio is 80, it means that it is willing to loan you 80% of the value of the house—providing, of course, that you meet certain credit requirements.

How much will a bank with a loan-to-value ratio of 80 loan you on a house that is worth $350,000? That requires you to take 80% of $350,000, which can be accomplished by the following succession of operations:

➤ Step 1: Enter 350,000 (there's no comma on a calculator, so you type in 350000. Enter means to type in just those numbers, so you type the six digits in 350000 and then go to step 2).

➤ Step 2: Press the x button

➤ Step 3: Enter 80

➤ Step 4: Press the % button

If you've done this correctly, the number 280000 will appear in the calculator window. The bank is willing to loan you $280,000—assuming you've got good credit.

As you'll see, this is also a template for doing the other types of percentage calculations which follow, such as adding a percentage to a number.

Adding a Percentage to a Number

The check arrives, and dinner has cost $40. You wish to add a 15% tip. Although your math teacher would hope that you could do this without a calculator (and you can do this fairly simply; just move the decimal point one position to the left to get $4, take half of that to get $2, and your tip is $4 + $2 = $6), here's the drill for doing it with a calculator:

> ➤ Step 1: Enter 40

> ➤ Step 2: Press the + button

> ➤ Step 3: Enter 15

> ➤ Step 4: Press the % button

If you've done this correctly, the number 46 will appear in the calculator window; that's the total amount that you have to pay. If the check has a separate line for a tip, you can compute the tip separately by taking 15% of $40 (as in the previous example). It's a safe guess that if you simply enter $46 on the total line, without filling in the amount of the tip, the restaurant personnel will fill it in for you when the time comes for them to complete the transaction.

Notice that the only difference between "taking a percentage" and "adding a percentage" is that in the former instance, you press the x button, and in the latter instance, you press the + button. You can probably guess what's coming next.

Subtract a Percentage from a Number

You've just made a purchase of $87.40 from a lumber yard and are in the process of taking out your credit card to pay for it. The clerk offers you a 2% discount for cash; a look in your wallet shows that you can cover it. To figure out how much you owe, go through the following (hopefully familiar) steps:

> ➤ Step 1: Enter 87.40

> ➤ Step 2: Press the - button

> ➤ Step 3: Enter 2

> ➤ Step 4: Press the % button

If you've done this correctly, the number 85.652 will appear in the calculator window; that's the total amount that you have to pay. You'll either end up paying $85.65 or $85.66, but the clerk is almost certain to be satisfied with $85.65.

Be Careful When Calculating Percentages!

This topic is very important, and percentages are a source of a great deal of confusion. Hopefully, this book will help eliminate some of that confusion—at least, for you. However, now that you've learned to compute percentages with a calculator, it's important to realize that the way a calculator computes percentages is not the same way we talk—or write—about percentages.

We always write "3% of 70." The number 3 comes first when we either write or discuss percentages, but the number 70 comes first when we use a calculator to compute percentages. The order in which you do things in mathematics and when using a calculator is extremely important. Although 3% of 70 and 70% of 3 are both equal to 2.1, adding 3% to 70 will give a result of 72.1, but adding 70% to 3 will give a result of 5.1. Big difference!

Calculators on Your Computer

If you have a computer, you actually have both a standard calculator and a scientific calculator. If you look under the Accessories tab in All Programs, you will see a calculator listed as one of the programs. The newer the software, the more options there are in the calculator. Early versions of Windows allowed you the choice of a standard or scientific calculator; Windows 7 gives you these two, as well as a programmer's calculator and a calculator for statistics. You choose which one to use by clicking on the View tab in the calculator and selecting the one you want to use.

There are advantages and disadvantages to using the calculator supplied with your software package (such as Windows). The Windows calculator lets you see the calculation you are performing, even if you perform a number of different operations in sequence, such as you might when adding up a column of figures. That way you can check to see that you've entered all the numbers—and arithmetic operations—correctly. The two drawbacks are that it's not so easy to stick your computer in your pocket—although they seem to be getting closer and closer to this every day—and that the computer's calculator doesn't always function in the same way that most handheld calculators do.

Here's an important example of this. The way the percent button works on the Windows standard calculator differs from the way it works on a standard calculator.

To either add a percentage to a number or subtract a percentage from a number, you go through the same four steps as above, but you must add a fifth (and mercifully final) step; pressing the equals (=) button. The Windows standard calculator will then display the correct answer. However, the Windows standard calculator displays the result of taking the percentage as an intermediate step. For example, if you were to add a 15% tip to a $40 meal, here's what would happen—and what the Windows standard calculator will display in its main window:

Main Windows Display

➤ Step 1: Enter 40   40

➤ Step 2: Press the + button   40

➤ Step 3: Enter 15   15

➤ Step 4: Press the % button   6

➤ Step 5: Press the = button   46

Notice that 15% of 40, which is 6, is displayed in the main window after step 4; a standard calculator that you actually hold in your hand would display 46.

It wouldn't matter whether you had pressed the +, x, -, or / button at step 2, what you see in the main window after step 4 will always be 6. So, if you want to take 15% of 40, you should note what is in the main window after step 4. What Windows does when you press the = button in step 5 is perform the operation specified by step 2 to the number that is in the main window after step 1 and the number that is in the Window after step 4. In this case, those two numbers are 40 and 6; so had you pressed the – button at step 2, the number 40 – 6 = 34 would be displayed in the main window after step 5. Had you pressed the x button at step 2, the number 40 x 6 = 240 would be displayed in the main window after step 5.

You can also find a variety of online calculators at the website MyCalculator.org, and you can generally find special-purpose online calculators simply by typing your needs into a search engine.

A Final Word of Advice Concerning Calculators

Try to have some idea of the size of the answer when you use a calculator. A calculator will not tell you when you've entered a wrong number or added when you should have subtracted. If your calculator tells you that you have more money in your checking account after you wrote a check than before you wrote the check, it's lying to you. Similarly, if you divide 81.23 by 17.418, this is roughly 80 divided by 20, or 4. It's actually about 4.66, reasonably close. If the calculator gives you an answer that is wildly different in size from what you expect the answer to be, the answer is quite likely to be wrong. Try doing it again.

Getting the Most Out of Your Cell Phone

The chances are pretty good that you actually have a calculator with you most of the time, because the chances are pretty good that you have a cell phone with you most of the time. Your cell phone comes equipped with a standard calculator; depending upon how your cell phone is structured, it's probably located under Office or Tools or some such name.

The newer smart phones have a scientific calculator available as an app. In fact, they often have a choice of several different models, if the word *model* can be applied to an app. Since apps are less expensive than physical calculators, you may want to consider buying a scientific calculator app.

The Language of Formulas

If you go to a foreign country, it's advisable to know a few basic words and phrases of the native language in order to make the trip a more pleasant one—you certainly don't want to walk into the wrong restroom. For many people, mathematics is like a foreign language, but if you master the mathematical equivalent of a few basic words and phrases, you'll be able to make very effective use of this book.

It is possible to learn how to use this book without ever learning the meaning behind what you are doing. There's no shame in that; most of us drive cars without having the foggiest idea of how an internal combustion engine works. However, it's worth spending a little time learning not only how to plug numbers into a calculator and press the buttons, but what the meaning of the arithmetic operations—the engine that powers the calculator—actually mean.

Haven't I Seen this Formula Before?

Every so often, you may look at a formula and think that it reminds you of another formula you've seen, just as a song may remind you of another song. There are only a limited number of notes and there are only a limited number of possible ways to write formulas.

Many formulas are produced through simple arguments relating to the common arithmetic operation. The questions, How long will it take me to diet to reach a certain weight? and How long will it take me before the value of my car depreciates to a particular amount? are essentially the same question—at least from the standpoint of the mathematics involved— and the formulas will look the same (although possibly different letters will be used).

People who are not very comfortable with mathematics may not always realize that one formula solves many problems. Even if they do, they may not always be able to make the transition from one situation to another. If you understand what *arithmetic operations* mean, you may be able to get even more mileage out of this book by realizing a formula that appears in one context actually applies in other contexts as well.

CHAPTER 2

 # Reading and Using Mathematical Formulas

In This Chapter

➤ Arithmetic expressions and operation

➤ How your calculator expresses large and small numbers

➤ Parentheses and PEMDAS

Can I Skip This Chapter?

You can decide for yourself whether to skip this chapter on the basis of the topics in the main headings. One such topic is how your calculator expresses large and small numbers, which requires something that is formally called scientific notation. If you know scientific notation, you can certainly skip that section. The chances are pretty good that you know everything you need to know about the basic arithmetic operations addition, subtraction, multiplication, and division, but you may need a quick refresher on exponents, parentheses, and PEMDAS.

Even if you decide to skip a section—or the entire chapter—you still own the book and can always come back to it if you feel there is something in that section you need to know.

Basic Arithmetic Operations: Addition

There are four basic arithmetic operations: addition, subtraction, multiplication, and division. They are the foundation of arithmetic, and it's worth devoting a little space to each of them.

Addition is the most fundamental of the arithmetic operations after counting, and it's not necessary to explain counting if you are reading this book. One way to describe addition is continued counting; if you have $10 and acquire another $3, you add 3 to 10 by continuing to count 3 additional numbers after 10. Ready? Here we go:

➤ 11 corresponds to the first of the 3 additional numbers

➤ 12 corresponds to the second of the 3 additional numbers

➤ 13 corresponds to the third of the 3 additional numbers

And there we are: adding 3 to 10 gives us a result of 13.

Adding Numbers on Your Calculator

The symbol for addition is the plus sign (+). It is the only symbol for addition. To add 10 and 3 on your calculator:

➤ Step 1: Enter 10

➤ Step 2: Press the + button

➤ Step 3: Enter 3

➤ Step 4: Press the = button

The number 13 will appear in the calculator window.

The order in which you add numbers does not affect the final result: 10 + 3 = 13 and 3 + 10 = 13 also.

Basic Arithmetic Operations: Subtraction

Most people think of subtraction as the opposite of addition; instead of continued counting, it's counting backward or taking away. If you have $10 and take away $3, you subtract 3 from 10 by counting backward 3 numbers from 10. Ready? Here we go:

➤ 9 corresponds to the first of the 3 numbers we take away

➤ 8 corresponds to the second of the 3 numbers we take away

➤ 7 corresponds to the third of the 3 numbers we take away

And there we are: subtracting 3 from 10 gives us a result of 7.

Subtracting Numbers on Your Calculator

The symbol for subtraction is the minus sign (-). It is the only symbol for subtraction. To subtract 3 from 10 on your calculator:

➤ Step 1: Enter 10

➤ Step 2: Press the - button

➤ Step 3: Enter 3

➤ Step 4: Press the = button

The number 7 will appear in the calculator window.

The order in which you subtract numbers does affect the final result: 10 - 3 = 7, but 3 - 10 = -7.

Zero and Negative Numbers

It's probably easiest to see positive numbers, zero, and negative numbers in the everyday context of your bank balance. A positive number such as 45.37 means you have money in your bank account—$45.37, to be precise. Zero means you have no money in your account; maybe you just closed it out. A negative number such as -23.15 means that you are overdrawn and actually owe the bank $23.15. A word of caution: all banks charge an additional sum—sometimes a hefty one—for being overdrawn. Arithmetic can often save you money. For instance, when you write a bunch of checks make sure that when you add up the total of all the checks, and then subtract that total from your bank balance, the result is a positive number.

Some negative numbers are meaningful; some aren't. Negative temperatures are meaningful; it means that it's really cold. As we've seen, negative bank balances are not only meaningful but ominous. However, we cannot talk about a negative number of refrigerators; although it makes sense to say "3 refrigerators" or "0 refrigerators," it just doesn't make sense to say "-5 refrigerators."

Basic Arithmetic Operations: Multiplication

Multiplication is shorthand for repeated addition of the same number. The number that results from multiplying 3 by 10 is the same as 10 + 10 + 10. A convenient way to think of this is the total number of items in 3 groups of 10 items each.

Multiplying Numbers on Your Calculator

There are two symbols for multiplication, and either might appear on your calculator. One is the times sign (x), and the other is the asterisk, or star (*). One or the other will appear on your calculator; we'll use the * in the example. To multiply 10 by 3 on your calculator:

➤ Step 1: Enter 10

➤ Step 2: Press the * button

➤ Step 3: Enter 3

➤ Step 4: Press the = button

The number 30 will appear in the calculator window.

The order in which you multiply numbers does not affect the final result: 10 * 3 = 30 and 3 * 10 = 30 also.

Basic Arithmetic Operations: Division

Probably the simplest way to think of division is as equal sharing. When we divide 12 by 3 we are sharing 12 equally among 3 groups. Division is the reverse (the technical word is *inverse* in case you ever come across it) operation of multiplication; in a sense, division undoes multiplication. If you multiply 3 by 4, you are counting the total number of items in 3 groups of 4. This number is 12. When you divide 12 by 3, you are separating 12 items into 3 groups of equal size. The result of dividing 12 by 3 is the number of items in each of these equally sized groups. So 12 divided by 3 is 4.

Dividing Numbers on Your Calculator

There are two symbols for division, and either might appear on your calculator. One is the old-time division sign ($\div$), and the other is the slash (/). One or the other will appear on your calculator; we'll use the / in the example. To divide 12 by 3 on your calculator:

➤ Step 1: Enter 12

➤ Step 2: Press the / button

➤ Step 3: Enter 3

➤ Step 4: Press the = button

The number 4 will appear in the calculator window.

The order in which you divide numbers does affect the final result: 12 / 3 = 4, but 3 / 12 = 0.25. Yes, it also equals ¼, but a calculator always displays numbers in decimal form.

Abbreviations and Formulas

You've already seen an example of both abbreviations and a formula in Chapter 1. That example was T = N x P, a shorthand for the total cost of buying a number of items at the same price. Letter abbreviations are common in mathematics and are somewhat misleadingly called variables, making you think that somehow they are going to vary, or change. Your bank balance is a good example of the same thing: in one sense it changes in that it has different values at different times, but in another sense it doesn't; it is always a bank balance, never a time of day or a temperature.

Expressions: Depositing $100 in Your Checking Account

An expression is just an instruction for performing arithmetic operations. We're not going to go into the rules for performing arithmetic operations here, because all you have to know to use this book is how to evaluate expressions by plugging numbers in to the appropriate formula.

Suppose that we let B stand for the current amount in your checking account, and let's suppose you deposit $100. Let A stand for the amount in your checking account after you deposit $100. Then B+100 is an expression that represents the result of adding $100 to whatever is in your checking account at the moment.

For instance, if there is $384.77 in your checking account, that's the value of B. When you plug in 384.77 to B, the expression B+100 becomes 384.77+100 = 484.77. That's the amount in your checking account after you deposit the $100, which we denoted by A. It doesn't matter what the current amount B is in your checking account; it is always true that when you deposit $100, A = B + 100. The formula A = B + 100 is just an abbreviation for the fact that whenever you deposit $100, your balance after the deposit is $100 more than your balance before the deposit, no matter what was in your account at the start.

In the previous paragraph, B + 100 is an expression that involves adding an actual number (100) to an abbreviation (B), but nothing mysterious is happening here. Even though the value of B + 100 is not known until somebody tells us what B is, it simply says that when you deposit $100 in your checking account, you will have $100 more, and that's always true.

One of the reasons that people have more problems with algebra than they really need to is that they think that something mysterious is going on that enables the addition of numbers to letters, and they were told when they were younger that you can't add apples and oranges. Whether or not you can add numbers to letters really doesn't matter (at least, not to you), because once someone tells you the value of B you can get out your calculator (if you need it) and add 100 to that value.

It's only a short step from that formula to realize that it is possible to replace the specific deposit $100 with a deposit D of any amount, resulting in the formula A = B + D. Of course, you use this formula all the time when balancing your checkbook, just as you use the formula A = B − C to compute your new balance after you've written a check in the amount C.

Multiplication Shorthand

There is a well-known type of shorthand that is useful for determining the product of two quantities when at least one of them is a variable. Instead of writing 3 x A or 3 *A, formulas will usually skip writing either symbol and simply write the two quantities next to each other as 3A. Similarly, instead of writing N x P or N * P, it saves space (important in long formulas) just to write NP.

Of course, you can't do this if there aren't any letter variables in the expression, for how would you know if 317 meant 317 or 3 x 17 x or 31 x 7?

When this convention is used and one of the quantities is a number, the number is written first. We write 3A rather than A3.

Exponents

Multiplication was invented as a type of shorthand for repeated addition of the same number, because back when people were scratching symbols on clay tablets, it was obviously a major headache to write down the sum of thirty-one separate 4s, for example.

Sometime later, it was realized that it would also be a good idea to have a type of shorthand for repeated multiplication of the same number. Thus was born the idea of exponents, and the convention that was adopted was to let the symbol 24 stand for the product of four 2s rather than write 2 x 2 x 2 x 2.

Some mathematical terminology is totally unimportant. For instance, it really isn't necessary to know which of the numbers in 10 - 3 is called the subtrahend (a number that is to be subtracted from a minuend) and which is called the minuend (a number from which the subtrahend is to be subtracted), because those terms never come up. However, the terms involved in the exponent operation do come up. In the expression 2^4, 2 is called the base and 4 is called the exponent. We sometimes use the term 2 to the 4th power as well.

We've reached the point where you probably ought to acquire a scientific calculator, because there are lots of problems that require being able to evaluate an expression involving exponents, and you just can't do that with a standard calculator. Fortunately, scientific calculators that you can put in your pocket are very cheap, and online scientific calculators are free.

Evaluating Exponents with a Scientific Calculator

There are lots of formulas, especially in financial calculations, that require computing exponents. Just as some calculators use x for multiplication and others use *, some calculators use the symbol y^x (or x^y) for exponents and others use ^. To calculate 2^4 on your calculator:

> ➤ Step 1: Enter 2

> ➤ Step 2: Press the exponent button (whichever of the above ones appear on your calculator)

> ➤ Step 3: Enter 4

> ➤ Step 4: Press the = button

The number 16 should appear in the main window of your calculator. The scientific calculator in Windows 7 also works like this (it has x^y as the exponent button).

Negative and Fractional Exponents

If we tried to interpret 2^{-3} in the same way that we interpreted 2^4 as the product of four 2s, we would run into trouble; what do we mean by the product of negative three 2s? Although we may not be able to interpret it; we can assign a value to it, which makes mathematical sense (you probably don't care how, so don't worry about it). We do so by calculating 2 to the power $-3 = 1$ divided by (2 to the 3rd power) $= 1/8$. Your calculator will evaluate $2^{-3} = 0.125$.

The steps are the same as outlined above except you have to do a little work to get a -3 in step 3. One way to do this is to replace step 3 with steps 3a, 3b, 3c, and 3d:

> ➤ Step 3a: Enter 3

> ➤ Step 3b: Press the (button (left parenthesis)

> ➤ Step 3c: Press the – button

> ➤ Step 3d: Press the) button (right parenthesis)

You probably just hate doing that. Rather than replace one step with four, there's a simpler way; just add a step 5.

> ➤ Step 5: Press the 1/x button (Most scientific calculators have one; for some you must press a second function button to activate it.)

This works because $y^{-x} = 1/y^x$; you might remember this from your math classes.

Similarly, if we tried to interpret $2^{0.25}$ in the same way, we would again run into trouble; what do we mean by the product of 0.25 twos? Again, we can assign a value to it, which makes mathematical sense. We do so by calculating $2^{0.25} = 2^{1/4} = 1.1892$.

Calculator Red Alert!

Your calculator will handle any exponent correctly as long as you are raising a positive number to some power. However, your calculator may not take certain negative powers of negative numbers. Trying to do so will generate an error message. Trouble usually arises if you are trying to take the square root of a negative number.

There is some good news. There's not much chance of this happening if you use a formula from this book and don't type an incorrect number into your calculator. That's because the problems in this book happen in the real world, with solutions that don't get into the problem of taking the square root of a negative number.

Large and Small Numbers: Scientific Notation

The decimal system is a wonderful thing. Any number you will ever need can be expressed as a decimal—and you can write it down. If you are a politician and want to talk about a program that will save two trillion seven hundred billion dollars over the next ten years, you can write 2,700,000,000,000. However, your pocket calculator can't; it can generally only write eight or nine digits in its main window. The scientific calculator in Windows 7 will enable you to write out numbers that are somewhat longer, but sooner or later it also runs out of space.

Scientific notation is simply expressing a number as the product of a number between 1 and 9.99999999999 (lots of 9s) and a power of 10.

Examples of Scientific Notation

Rather than go into lengthy explanations, here are a few examples:

Number	Scientific Notation	Calculator Display
2,700,000,000,000	2.7×10^{12}	2.7 E12
0.000000000123	1.23×10^{-10}	1.23 E-10

There is some variation in how the calculator displays the number, but it will be similar to the above examples.

Parentheses and PEMDAS

There are some things you can do where the order in which you do them doesn't matter. If you put one squirt of cream and two spoonsful of sugar in your coffee, you can put the cream in first and then the sugar, or vice versa, and the coffee will taste the same. We've seen that the order in which you add or multiply two numbers doesn't matter, either.

On the other hand, there are some things you can do where the order in which you do them will make a great deal of difference. If you are away from home, it almost certainly will matter if you decide to drive home first and undress to take a shower, or whether you undress first and then drive home. Subtraction, division, and exponentiation are examples of mathematical operations in which the order does make a difference.

So far we've only discussed operations involving two numbers. We could certainly describe a sequence of operations involving three numbers; we could say, for instance, What is 1 + 2 × 3? It sounds straightforward enough, but hidden in this question is an ambiguity. Do we mean 1 + 2 pause × 3, in which case we would add 1 + 2 getting 3, and then multiply the result by 3 to get 9? Or do we mean 1 + pause 2 × 3, in which case we would multiply 2 × 3 getting 6, and then add this to 1 to get 7?

It's a genuine ambiguity, and mathematics resolves it by using parentheses—and an acronym called PEMDAS.

What Parentheses Tell Us to Do

It's simple: parentheses have priority over all other operations.

Let's look at how parentheses resolve the above ambiguity.

➤ (1 + 2) x 3 = 3 x 3 = 9 (the parentheses made us add 1 + 2 before performing the multiplication)

➤ 1 + (2 x 3) = 1 + 6 = 7 (the parentheses made us multiply 2x3 before performing the multiplication)

➤ In case of multiple pairs of parentheses, we work from the innermost pair of parentheses outward. For example:

1 + (2 x (3 – (4 x 5)))= 1 + (2 x (3 – 20))

= 1 + (2 x-17)

= 1 + -34

= -33

PEMDAS: Please Excuse My Dear Aunt Sally

When you see a red traffic light, you stop. The color red does not intrinsically mean stop—it's a convention on which people have agreed.

In order to eliminate unnecessary parentheses, mathematicians have agreed on a convention that eliminates unnecessary parentheses. This convention is known as PEMDAS. Students remember it by the first letters of Please Excuse My Dear Aunt Sally. Expressions in parentheses (P) are evaluated first, then exponents (E), then multiplications (M), then divisions (D), then additions (A), and finally subtractions (S).

Generally, the P and E of PEMDAS are clear—you can see what's inside the parentheses, and exponents are always raised and to the right, as in 24.

PEMDAS in action

Suppose that you are asked to evaluate 18 – 2 + 6 * 4 / 2:

➤ 18 – 2 + 24 / 2   (M)

➤ 18 – 2 + 12   (D)

➤ 18 – 14   (A)

➤ 4   (S)

This could also have been written 18 – (2 + ((6 * 4)/2)) to eliminate any ambiguity by using parentheses rather than PEMDAS. Although there is nothing wrong with doing this, it's a little like wrapping a present in several different layers of wrapping paper. It doesn't hurt anything, but it just makes for more work than necessary.

Familiarity with PEMDAS makes it possible to write down formulas without using parentheses unnecessarily. Hopefully, all the formulas you will see here do not use parentheses unnecessarily to make for shorter formulas that are easier to read.

CHAPTER 3

Formulas: Beyond the Basics

In This Chapter

➤ Special functions

➤ Mathematical models and unstated assumptions

➤ Introduction to Excel (or Numbers for Mac users)

Can I Skip This Chapter?

You should probably skip this chapter only if you are comfortable with exponential, logarithmic, and trigonometric functions, and are familiar with how to evaluate these functions using a scientific calculator. You can also skip this chapter if you know for certain the only problems you would ever encounter could be solved by a standard calculator—but that's very unlikely.

Many people already have a working knowledge of Excel. If you use Excel even on an occasional basis—enough so that you know what the equals sign (=) does in Excel—you can skip the section on Excel. If you purchased this book only as a reference for occasional problems that you encounter, you can skip the section on Excel. Certainly, if you don't have access to a computer that has Excel on it, there's absolutely no reason to find one because you've gotten along quite comfortably without it.

However, you may find that there are problems in the book that you use on a recurrent basis. Maybe you are entrusted with using recipes from a cookbook and continually have to figure out exactly how many cups of sugar to use because the recipe is worked out for four people, and you continually find yourself making it for more or fewer than four people. If you frequently need to make the same calculation with a different collection of numbers, it is worth spending the time it takes to read the Excel section.

Calculators or Spreadsheets?

There's a big problem with calculators that is familiar to all teachers in courses that use them. Students believe what their calculator says and are frequently unaware when they have either entered a number erroneously or pushed the wrong button. Of course, it is always possible to check a calculator computation by redoing it; if you get the same answer, it is very likely to be correct. However, if your answers differ, you are going to have to do the calculation a third time to see which answer was correct.

In addition, calculators require you to press various buttons to perform arithmetic operations as well as enter the numbers from the calculation. When a formula is long, this is often a tedious process with a significant chance of error. However, once you have correctly entered the formula into a spreadsheet, there is virtually no chance of error. All you have to do is type the numbers in the formula in the correct location, and the formula will be correctly computed using those numbers. The only error that can be made is entering the wrong number, but you can see the number that you entered because it stays precisely where you typed it. Some of the more expensive calculators display the computation, allowing you to see all the numbers you enter as well as the operations. However, the cheap ones simply do what you tell them to do when you tell them to do it, displaying only the result of the last key press.

If it doesn't cost you anything except a little time, give Excel (or Numbers) a try. Once you have, you'll never go back.

One other thing: If you have a computer but haven't bought an office package, check out Open Office. It has free word processing and spreadsheet programs, plus a whole lot more. It's also compatible with Microsoft.

Special Functions

There are three important functions that are generally not found on scientific calculators, although they do appear in spreadsheets (another minor reason for using spreadsheets).

The Absolute Value, Maximum, and Minimum Functions

The absolute value of a number x, written $|x|$, is simply the size of the number. If x is positive or 0, $|x| = x$. If x is negative, such as -4, its absolute value is found simply by removing the minus sign; $|-4| = 4$.

The maximum of a collection of numbers is the largest of those numbers. The largest of the numbers 3, 4, 8, 2, 8, 1 is 8; using the maximum function, this would be written max(3, 4, 8, 2, 8, 1) = 8. Notice that the same number can appear twice—or more—in a list, and that the order in which we write the numbers down really doesn't matter.

This probably won't come as a big shock: the minimum of a collection of numbers is the smallest of those numbers. Using the same collection of numbers and the obvious abbreviation, we have min(3, 4, 8, 2, 8, 1) = 1.

If you have to find the maximum or minimum of a short list of numbers, such as the ones above, you don't need a calculator, you can just eyeball it. However, if you have a really long list of numbers, there's a fabulous way to do this in Excel, which we'll discuss later.

Scientific Calculators: Vive la Différence!

Scientific calculators not only differ greatly from standard calculators, they differ greatly from each other. The higher-end scientific calculators come with graphing features that would have been the envy of scientists and engineers as little as fifty years ago—and they cost less than the price of a meal for two at a good restaurant.

There is a common aspect to many scientific calculators: almost all have buttons for the trigonometric functions (more on that later), but the inverse functions (more on that later as well) and other features vary widely. Sometimes there is a specific button (which can go by different names on different calculators), and sometimes one needs to activate what is called a 2nd function button as well. The best thing to do is to consult the website of your calculator manufacturer for specifics.

The Factorial Function n!

This function comes up a lot in situations involving probability and statistics. It is only defined for nonnegative integers (whole numbers). If n is a nonnegative whole number, the expression n! is read as "n factorial"—unless you happen to be in England, where it is read as "n shriek," because that's what they call exclamation points.

0! is simply defined to be equal to 1. If n is a positive integer, n! is the product of the positive integers from 1 through n. For example, 5! = 1 x 2 x 3 x 4 x 5 = 120.

Most scientific calculators have a button for this function, but you will get an error message if you have anything other than a nonnegative whole number in the main window of the calculator when you press it. Factorials get very large very quickly. The number of different ways that a deck of cards can be shuffled can be expressed as 52!, which is about 8.07 x 1067, so if you are going to run into problems that use the factorial function, it's a good idea to be comfortable reading how your calculator displays large numbers.

Exponential and Logarithmic Functions

Even though there is a button for computing any legitimate power of any base (*legitimate* here means that you won't get an error message, which you do if you try to take the square root of a negative number), there are two special bases that are used so frequently that they have been deemed worthy of buttons all their own. These bases are 10, which is certainly familiar to you, and e, which may not be.

e *Doesn't Always Mean "Electronic"*

There are two numbers that play a critical role in practical mathematics. Most people are familiar with π, which is approximately equal to 3.14159 and is the quotient of the circumference of a circle divided by its diameter. Of equal importance is the number e, which is approximately equal to 2.71828.

The number e appears naturally in connection with growth and decay processes, where the rate at which something grows or decays is proportional to the amount of stuff that is either growing or decaying. If you have twice as many rabbits, you get twice as many bunnies. It also shows up in a number of areas such as finance and probability.

The 10x button and the e^x button function in exactly the same way: when you press the button, it raises 10 (or e) to whatever power is in the main window at the time you press the button. Here are a couple of examples you can check out by simply entering 2 in the main window:

Button Name	Example
10^x	$10^2 = 100$

e^x $e^2 = 7.389\ldots$ (the three dots indicate that some numbers that appear on your calculator have been left out)

You don't need to remember all that stuff about logarithms that you may have forgotten. For those who haven't forgotten, the log button takes logarithms to the base 10 (these are called common logarithms), and the ln button takes logarithms to the base e (these are called natural logarithms). Here are two more examples you can check by entering 100 in the main window:

Button Name	Example
log	$\log 100 = 2$
ln	$\ln 100 = 4.605\ldots$

Logarithm Lookouts

Logarithms have something in common with square roots: the logarithms of negative numbers are not defined. Try taking a logarithm of a negative number and you'll get an error message. The logarithm of 0 is also not defined. Of course, these shouldn't cause you any problem in the real world because a real-world solution to a problem won't involve taking a logarithm of a negative number or a logarithm of 0.

Logarithms are functions, and the logarithm of a complicated expression should always have parentheses around it to indicate exactly which expression is having its logarithm computed. To illustrate: log 2 = .30103 (to 5 decimal places), and log 10 = 1. log(2 + 8)= log 10 = 1 (remember the P in PEMDAS), but log 2 + 8 = .30103 + 8 = 8,30103.

Angles and Trigonometric Functions

Most calculators have three different ways to measure angles: degrees, radians, and grads. You can skip grads; they are rarely if ever used. Degrees are the common way of measuring angles. There are 90 degrees in a right angle and 180 degrees in a straight angle.

Radians are a natural way to measure angles in the sense that measuring angles in radians prevents the appearance of "ugly" constants such as $\pi/180$. However, most people measure angles in degrees, and the formulas that appear in this book generally assume that angles are measured in degrees—unless specifically stated otherwise.

You don't have to worry, but you do have to make sure that your calculator is using the correct system of measuring angles, which is generally degrees. One way to do this is to check with the following evaluations of the trigonometric functions of sine (the sin button), cosine (the cos button), and tangent (the tan button):

Button Name	Example
sin x	sin 30 = 0.5
cos x	cos 60 = 0.5
tan x	tan 45 = 1

If your calculator is erroneously set to radians or grads, you won't get 0.5 when you compute sin 30.

Because sin, cos, and tan are functions, care must again be exercised by using parentheses to distinguish between sin(30 + 60) and sin 30 + 60. Parentheses always come first, so sin(30 + 60) = sin 90 = 1 (as long as your calculator is set to degrees), but sin 30 + 60 = 0.5 + 60 = 60.5.

Inverse Trigonometric Functions

In a nutshell, inverse trigonometric functions play the same role for trigonometric functions that the square root function does for the square function. Squaring a positive number, and then taking its square root, gets you back to the original number; the square root of 52 is 5. Within certain limitations, starting with an angle, taking its sine, and then taking the inverse sine of that number gets you back to the original number:

Button Name   Example

$\sin^{-1} x$ $\sin^{-1} 0.5 = 30$

$\cos^{-1} x$ $\cos^{-1} 0.5 = 60$

$\tan^{-1} x$ $\tan^{-1} 1 = 45$

Inverse Trigonometric Functions: A Word of Caution

Your scientific calculator wants to make you happy, but it can go a little too far when dealing with inverse trigonometric functions. The expression $\sin^{-1} 0.5$ is the angle whose sine is equal to 0.5. This angle is either 30 degrees, if you're measuring angles in degrees, or $\pi/6$ = 0.5236 radians. Your calculator will show you whatever you want to see, 30 or 0.5236 , depending upon whether you've set the angle mode to degrees or radians.

That's fine if all you want to do is find the angle measure of a particular angle. However, if the formula you are using involves further computations involving $\sin^{-1} 0.5$, you must make sure that the calculator is set to radians, as all scientific, mathematical, and engineering formulas assume that angles are measured in radians.

This is another reason to learn to use a spreadsheet, as spreadsheets generally evaluate inverse trigonometric functions in radians, and you won't run into any problems.

Subscripts

Lists are an important way of writing down information that is somehow related, such as a list of errands to be performed or a grocery list. Lists of numbers occur frequently, such as a list made of the total cash receipts of a business on a daily basis. We generally refer to specific items in a list by giving its position in the list. For instance, suppose we have the following list of numbers:

5, 7, 8, 2, 14, 9, 35.

The number 14 is the fifth item on the list. Mathematics emphasizes consistent ways of referring to things, so a typical approach to this problem is to use subscripts. A general list of numbers could be defined using this approach by saying:

➤ X_1 = first number on list

> ➤ X_2 = second number on list

> ➤ N = how many numbers are on list

> ➤ X_N = last number on list

In the above example, N = 7, and X_5 = 14.

If we wish to refer in some way to a consecutive group of items from a list, we use three dots to indicate the missing items. For instance, we might refer to the whole list of all the numbers from X_1 through X_N by using the notation $X_1, X_2, ..., X_N$. If we needed to refer to the sum of the numbers from X_2 through X_{14}, we would do this by writing $X_2 + X_3 + ... + X_{14}$.

Mathematical Models and Unstated Assumptions

There's probably a little bit of Buyer Beware! in every book—more if you read a book on how to treat a heart condition or how to buy stocks, less if you read a book like this. Some of the formulas in this book are always true, such as the total cost of a number of items at the same price being the product of the cost per item and the number of items. Some aren't, either because a mathematical model of a real situation was used to get an approximate answer for a problem, or because there is an unstated assumption (or several of them) that would be so clumsy or time-consuming to state that they just aren't stated.

Mathematical Models

A mathematical model is often an attempt to simplify a complicated real-world situation in order to make it amenable to mathematical analysis. One such example is the calculation of how much time it takes fans to reduce the smoke level in a smoky room. This model assumes that the smoke particles are well mixed and that each cubic foot of air contains the same number of smoke particles. Obviously, that isn't completely true, but it's probably close to the actual situation and, more importantly, allows us to construct a mathematical model that comes up with a useful result.

Some of the models used require substantial knowledge to construct, and there are often several alternative models available. When such a model is used, it is one that has attained widespread acceptance and has been proven useful in the real world.

An Unstated Assumption

Many problems make the assumption that a certain number remains constant throughout the duration of the problem. A problem that states that a person can rake the leaves in 2 hours tacitly assumes that this rate remains the same during the entire problem, and can be extrapolated to remain the same whenever that person is raking leaves. He will rake half as many leaves in 1 hour, or twice as many leaves in 4 hours. While this assumption is almost certainly not perfectly true, it is true as an approximation and is often the only way to approach a problem.

You're Ready to Make Use of This Book!

You now have all that you need to use this book. If you learned or reviewed a little math along the way, hopefully it wasn't too painful an experience. If you find later in the book that you can't read a formula, you can always come back to either Chapter 2 or this chapter, Chapter 3, to find what you need.

However, you can make more effective use of the book if you read the next section on Excel. It's only a few pages, and it will save you a lot of time and prevent miscalculations as well.

Excel and Other Spreadsheets

You may find yourself using some of the formulas in the book frequently, and some of them are a little on the lengthy side. If you use a calculator, you may have to keep track of intermediate results, and this can be time-consuming and increases the chance of error. It is very easy to acquire enough proficiency with a spreadsheet such as Excel to make these computations much more quickly and accurately.

What follows is a relatively simple tutorial, which will give you more than enough proficiency with Excel to make frequent computations extremely quickly.

A Spreadsheet at Startup

When you open an Excel worksheet (either an .xls file for the older versions of Excel or an .xlsx file for the newer versions), you find yourself confronting a table with rows and columns. The rows are labeled 1, 2, 3… on the left of the screen. The columns are labeled A, B, C,… across the top of the screen. The expression B2 is not a vitamin but refers to the cell in column B, row 2.

	A	B	C
1			
2			
3			
4			

There are some toolbars above this on the worksheet, but you needn't worry about that at this moment.

Data Entry

Data entry in Excel is simple: move the cursor (indicated by an outlined white plus sign) to a particular cell and click the mouse. The cell will be outlined in black. It's also helpful that the row designations at the left of the sheet and the column designation at the top of the sheet for that cell will change from gray to yellow. This is useful if you should click on a cell such as J241, as it can be hard to tell whether you've clicked on J241 or a neighboring cell such as J240 or K241.

Now is the time to look at the bar right above the column letters. You will see that it has a short white space at the left, then a shorter gray space with the notation fx in it (the f is in script), and finally a long white space. If you move the mouse to cell C3 and click it, C3 appears in the short white space.

To enter text, simply start typing. When you have finished, press the Enter or Return key on your keyboard. The text will appear in the cell and in the long white space on the bar above the column letters.

To enter numbers, do exactly the same thing. If you enter a number, Excel will treat it as a number so you can do arithmetic with it.

You may want to adjust the width of a cell to allow for a longer text entry. You can do so for all the cells in column B, for example, by positioning the cursor on the vertical bar that separates the letter B from the letter C in the previous example. Now hold down the left mouse button and move the mouse to the right. You will see the cells in column B widen as you do so. When the cells are at an acceptable width, release the left mouse button.

In the example below, column B has been widened so the text entry Total Cost can be displayed in cell B2.

	A	B	C
1			
2		Total Cost	
3			
4			

Performing Computations: The Big Reason for Using Excel

First things first. Excel uses + for addition, - for subtraction, * for multiplication, / for division, and ^ for exponentiation.

Let's say that you have a spreadsheet that looks like this:

	A	B	C
1			
2	3	5	
3			
4			

You would like the contents of cell C2 to be the result of adding 6 to the product of whatever numbers are in A2 and B2. At this moment, 3 is in A2 and 5 is in B2, but you may decide to change them later.

The Excel expression for adding 6 to the product of whatever numbers are in A2 and B2 is either 6+A2*B2 or A2*B2+6. Yes, Excel knows about PEMDAS! No harm will come if you use 6+(A2*B2), but it's not necessary. No harm will come if you use lowercase letters (entering a2 and b2), either.

To perform the above computation and have the result in cell C2, move the mouse to cell C2 and click on it. Then start the data entry with an equals sign, and use either numbers or cell references along with Excel's arithmetical operators. If you enter the expression

=A2*B2+6

into cell C2, Excel does exactly what you want it to. You will notice that the spreadsheet now looks like this

	A	B	C
1			
2	3	5	21
3			
4			

One of the most important features of Excel is automatic recomputation. Should you go back to cell A2 and change its value to 4, the moment you press the Enter key on your computer keyboard, the spreadsheet will instantly change and look like this:

	A	B	C
1			
2	4	5	26
3			

4			

Putting a Formula into Excel

Let's see how to implement the formula

total cost = number of items x cost per item

into an Excel spreadsheet. While there are numerous ways to do this, here's a very straightforward way:

Start by typing text entries into cells A1 through D1 so the spreadsheet looks like this:

	A	B	C	D
1	**Total Cost Formula**	**Cost Per Item**	**Number of Items**	**Total**
2				
3				
4				
5				

Into cell D2 type the expression

=B2*C2

After you have pressed the Return key to implement this, the spreadsheet will look exactly like the above spreadsheet, but it will have a 0 in cell D2. This shouldn't surprise you, after all you haven't bought anything (Excel interprets a blank in B2 as having the numerical value 0) and you didn't pay anything for it, either (similarly, Excel interprets a blank in C2 as having the numerical value 0).

OK, let's buy something. Let's buy 3 pounds of apples at 70¢ per pound. We can either leave cell A2 blank, or we can put in a description of the purchase, as in the following spreadsheet:

	A	B	C	D
1	**Total Cost Formula**	**Cost Per Item**	**Number of Items**	**Total**
2	3 pounds apples	0.7	3	2.1
3				
4				
5				

You actually typed in 0.70 in cell B2, but Excel overrode you and simplified the number. You know that D actually represents $2.10, but Excel doesn't, and so it prints out 2.1 instead of $2.10.

You'd probably prefer to see 70¢ in cell B2 and $2.10 in cell D2. This can be arranged using the Format feature of Excel, but you'll have to go into that on your own. It isn't hard, but nice-looking formatting is not what this book is about.

It's ultra-important that you save that spreadsheet! Absolutely nothing is more annoying than to put in time and effort doing something and then losing it. You've put in time and effort to create a spreadsheet—be sure you save it! When you open the spreadsheet again, it will appear exactly the same way you left it.

Special Feature #1: Fill

Let's look at that last spreadsheet. It occurs to you that it might be nice to be able to compute the total price for three different purchases and add them up. There is a simple and tedious way to do this. Enter:

> ➤ =B3*C3   in cell D3

> ➤ =B4*C4   in cell D4

> ➤ Grand total   in cell A5

> ➤ =D2+D3+D4   in cell D5

This doesn't take too much time, but imagine if you were a large store buying hundreds of different items. Doing it this way would take the better part of a day.

What you want is a quick way to tell Excel that you want rows 3 and 4 to mirror row 2; to take the product of cell B in a particular row, multiply it by cell C in the same row, and put the result in cell D of the same row. The Fill command in Excel accomplishes just that!

Position the mouse over cell D2 and left click but hold down the left mouse button. Then move the mouse down to cell D4 and release the left mouse button. You will notice that cell D2 is colored white, but the remaining cells (D3 and D4) will be colored blue.

Now look for Fill: in Excel 2007 it's a button at the top right of the screen, in Excel 2003 you can find it on the Edit menu, and if you have an earlier version of Excel it's time to upgrade. If you click the Fill button (in Excel 2007) or go to the drop-down menu (in Excel 2003), you'll find that one of the options is Down. Click it. Done!

This takes about the same amount of time and effort as simply typing =B3*C3 in cell D3 and =B4*C4 in cell D4, but it takes a lot less effort if you want to do it for five hundred more rows rather than just two.

Special Feature #2: The Colon (:)

The colon is a special character in Excel enabling one to deal with large quantities of data, as long as that data is in rows or columns.

Using the Colon and Worksheet Functions

The contents of cells A4 through A17 can be treated as a single item by typing A4:A17. To compute the sum of these items and locate it in cell B5 (for whatever reason), simply type

=sum(A4:A17) in cell B5.

This is an example of a worksheet function; the Sum command adds whatever you ask it to add as long as you separate the items to be added by commas. For instance, the expression sum(A2,A4:A17,43) adds the following numbers together: the contents of cell A2, the contents of cells A4 through A17, and the number 43.

Other extremely useful worksheet functions (Excel has gazillions) are max and min. They work exactly like sum; max finds the largest of a collection of items separated by commas, and min finds the smallest.

Colons can be used to refer to cells in a row, such as the contents of cells C5 through G5, which is abbreviated C5:G5.

Colons can even be used to refer to all the cells in a rectangular portion of a worksheet. The expression C3:G7 refers to that portion of the worksheet in rows 3 through 7 inclusively and columns C through G inclusively.

A Grocery List Spreadsheet

Let's fill in a little more info in our spreadsheet. It might end up looking like this:

	A	B	C	D
	Total Cost Formula	**Cost Per Item**	**Number of Items**	**Total**
1				
2	3 pounds apples	0.7	3	2.1
3	5 pounds hamburger	1.29	5	6.45
4	2 frozen pizzas	4.49	2	8.98
5	Grand Total			17.53

Not pretty, but effective. How could you pretty it up? Learn how to format numbers, so that $2.10 would appear in cell D2 rather than 2.1. You might also start your individual items in row 3, leaving row 2 blank, and then leave row 6 blank as well, using row 7 for the Grand

Total row. It takes up a little more space but is easier to read.

If you really want to learn about Excel, there are plenty of free tutorials online, or you can buy a book.

The Principal Benefit of Spreadsheets

Once you've verified that a formula is correct, you never need to compute that formula again. In the example above, once you have typed—and verified—that row 2 is correct, you can use it over and over again simply by typing new item costs and number purchased in cells B2 and C2 respectively.

So how do you verify that you've typed a formula in correctly? Formulas in this book come with a solved problem. That problem does two things for you. First, it helps you decide whether this is the problem you want to solve. Second, the fact that there is a verified solution enables you to check to see that you've entered the formula correctly by using the numbers in the problem. If your answer agrees with the one in the book, you've typed the formula correctly!

It requires a little bit of work to become more familiar with Excel, but you will want to learn about worksheet functions in order to be able to evaluate trigonometric functions and logarithms. Excel is a powerful and sophisticated tool. Its utility extends far beyond what has been outlined here, but a small amount of effort spent learning how to enter and evaluate expressions such as are found in this book will pay enormous dividends.

PART TWO

Finances

 Buying and Selling: Percentages

In This Chapter

➤ Buying low and selling high

➤ Markup and markdown percentages

➤ Common percentage misunderstandings

Can I Skip This Chapter?

We spend a fair amount of time on a typical day buying and selling things. Practical math—both the subject and this book—spends a lot of time analyzing this process. Benjamin Franklin once said that a penny saved is a penny earned. Even if that isn't literally true, everyone realizes how important it is to save money. After all, when you pay too much for something, it's money out of your pocket.

Buying and selling involve everyday financial decisions. Most people will face the problems in this chapter with some frequency. Additionally, many people are insecure about percentage computations, and if you are one of those people, you will benefit from this chapter.

Purchasing in Bulk

This is one of the most common purchasing decisions we face. You can buy practically everything—food, toothpaste, detergent—in an assortment of sizes. How much money can you save by buying something in a larger size?

Practical Math Problem

4.1 Buying in Bulk

A small box of cornflakes contains 12 ounces and costs $2.00. A giant box of cornflakes contains 18 ounces and costs $2.50. How much do you save by buying the giant box?

➤ c = cost of small package

➤ n = number of items in small package

➤ C = cost of large package

➤ N = number of items in large package

➤ S = amount saved buying large package

$$S = \frac{Nc}{n} - C$$

You will save 50¢ by buying the larger package.

These were the actual prices for two boxes of cornflakes on the shelf of a local grocery store at the end of 2010. Saving 50¢ may not seem like a big deal, but do this over and over again and you'll find yourself saving hundreds, possibly even thousands, of dollars a year.

However, in order for the savings to actually occur, you have to make sure that you consume the product! Mayonnaise, for instance, can spoil, and maybe you really don't use a whole lot of mayonnaise. If that's the case, you're actually going to be wasting money if you buy the large jar and have to throw most of the mayonnaise out. But cornflakes and detergent and a whole lot of other products last nearly forever.

Insurance Policy Deductibles

This innocuous-seeming question can be worth tens of thousands of dollars over the course of a lifetime. Insurance is an important part of our lives: we are not allowed to drive unless we have automobile insurance, most of us have—or hope to have—health insurance, and anyone who owns a home or a business has insurance to guard them against catastrophic loss.

A key feature of every insurance policy is the amount of the deductible. If you have a $100 deductible automobile insurance policy and you have an accident, once the insurance company approves the claim, it will send you a check for the amount of the repair less $100, the deductible amount. Many policies offer you the choice of several different deductible amounts. The one you should choose depends upon how likely you think you are to have an accident. As a rule of thumb, the less likely you think you are to have an accident, the higher the deductible amount you should choose. This is because the premium the insurance company charges for higher deductibles is less than the premium it charges for lower ones, as the company has to write smaller checks when the person who buys the policy chooses the higher deductible.

Practical Math Problem

4.2 Choosing a Deductible

An automobile insurance policy with a $100 deductible has a premium of $240. The same policy with a $300 deductible has a premium of $200. What percentage of the time must a claim be filed in order to make the policy with the higher premium a better buy?

➤ d = smaller deductible amount

➤ p = premium associated with smaller deductible

➤ D = larger deductible amount

➤ P = premium associated with larger deductible

➤ r = percentage of time claim needs to be filed

$$r > \frac{100(p-P)}{D-d}$$

You would have to exercise more than 20% of the policies to make the one with the higher premium—and the lower deductible—a better buy.

This problem shows the importance of being able to estimate how frequently something will happen, and to be able to translate that into a percentage. If you need to exercise more than 20% of the policies, let's face it, you're a lousy driver. Because the past is a pretty good indicator of the future, you can look at your own driving record to help you make the decision.

And if you're a business owner or a home owner, there are tables of percentages published by insurance companies that will help you make the decision as well.

Choosing a Sales Job

Many sales jobs offer a base salary plus commission on sales. Although a firm may sometimes offer a prospective employee the choice of two possible base-plus-commission plans, a more likely possibility is that two different firms will offer the same prospective employee two different plans.

Practical Math Problem

4.3 Choosing a Commission Plan

A salesman is offered a choice of two plans: a base salary of $25,000 plus 5% commission on annual sales, or a base salary of $20,000 plus 7% commission on annual sales. How much must he sell annually in order to make more money by accepting the $20,000 plus 7% commission arrangement?

> ➤ S = higher base salary

> ➤ P = commission percentage associated with higher base

> ➤ s = lower base salary

> ➤ p = commission percentage associated with lower base

> ➤ B = break-even sales total

$$B = \frac{100(S - s)}{p - P}$$

The salesman must sell more than $250,000 annually in order to make more money with the lower base salary and higher commission.

This problem is a very good example of the role of unstated assumptions. If the salesman is offered a choice of the two plans at the same company, the formula makes the situation very clear. However, there are complications if the salesman is offered this choice at two different companies; it might be much easier to sell $250,000 of one product than of another.

Finally, this problem can occur in situations other than sales. An investment manager was recently confronted by the following situation. He was offered the opportunity to manage a fund for 20% of all profits, or 25% of the first $100,000 of profits and 15% of all profits over $100,000. It's not quite the same problem, but it's certainly similar.

Monthly Average Needed to Make a Sales Goal

Typically, salesmen—or saleswomen—know exactly how much they need to sell during a given year to reach a sales goal. They will generally divide this figure by 12 in order to get a monthly sales goal, so then they will have some idea of how they are doing on a month-by-month basis. Sometimes they exceed the goal, sometimes they fall short. Once several months have gone by, it is a good idea to readjust the monthly average to see what it should be to meet the annual goal.

Practical Math Problem

4.4 Reaching a Sales Goal

A salesman makes 5% commission on all his sales. He has made $800,000 in sales through the end of July. How much must he average in monthly sales in order to make $100,000 in commissions for the year?

➤ S = sales to date

➤ m = number of months left in the year

➤ p = commission expressed as a percentage

➤ Y = desired yearly commissions

➤ A = average monthly sales needed

$$A = \frac{\dfrac{100Y}{p} - S}{m}$$

The salesman must average $240,000 a month in sales in order to meet his goal.

This problem appears several times in this book in different forms. One not-so-obvious version of this problem occurs on election night, when a known percentage of the returns are in, and a candidate has a certain percentage of the vote. The sales goal here is 50% of the total vote, and the average sought is the percentage of the remaining vote that is needed to put the candidate over the top.

Choosing the Best Allocation of Assets

This next problem has a complicated solution involving separate steps, but it's an extremely important one. Whenever we allocate assets, we want to do so in a way that generates the most "bang for the buck." This problem will be presented in a venue that is probably familiar to everyone, and afterward several different variations of the problem will be discussed so you can see how widely applicable the result is.

Practical Math Problem

4.5 Best Use of Two Ingredients

A company manufactures both a high-protein and a high-carbohydrate trail mix. One pound of the high-protein trail mix uses ¾ of a pound of nuts and ¼ of a pound of fruit, and sells for $4. One pound of the high-carbohydrate trail mix uses 5/8 of a pound of fruit and 3/8 of a pound of nuts, and sells for $3. The company has 300 pounds of nuts and 240 pounds of fruit on hand. How many pounds of each type of trail mix should it make to maximize its revenue?

In problems such as these, it makes it easier to read if the information is presented in table form. Here's one way to do it:

Data for 1 Pound of Each Product

	Fruit	Nuts Sale	Price
High Protein	0.25	0.75	4
High Carbohydrate	0.625	0.375	3
Pounds Available	240	300	

If we label fruit and nuts as Items, and the two types of trail mixes as Products, we can construct a generic version of this table as follows:

Data for 1 Unit of Product

	Item 1	Item 2	Sale Price
Product 1	a_1	b_1	c_1
Product 2	a_2	b_2	c_2
Units Available	A	B	

The solution to the problem will be:

➤ X = number of units of Product 1

➤ Y = number of units of Product 2

➤ M = maximum revenue

We now compute the maximum revenue M, which will be the smallest of the numbers P, Q, and R.

$$P = c_1 \min\left(\frac{A}{a_1}, \frac{B}{b_1}\right)$$

$$Q = c_2 \min\left(\frac{A}{a_2}, \frac{B}{b_2}\right)$$

$$R = \frac{c_1(Ab_2 - Ba_2) + c_2(Ba_1 - Ab_1)}{a_1 b_2 - a_2 b_1}$$

$$M = \max(P, Q, R)$$

$M = \max(P, Q, R)$

You've now determined what the maximum revenue M will be. The next step is to determine X, the number of units of Product 1 you must produce, and Y, the number of units of Product 2.

If M = P, you won't need to produce any units of Product 2.

$$X = \min\left(\frac{A}{a_1}, \frac{B}{b_1}\right) \quad Y = 0$$

$Y = 0$

If M = Q, you won't need to produce any units of Product 1.

$$X = 0 \quad Y = \min\left(\frac{A}{a_2}, \frac{B}{b_2}\right)$$

$X = 0$

Finally, if M = R, you're going to have to produce some of each.

$$X = \frac{Ab_2 - Ba_2}{a_1b_2 - a_2b_1} \quad Y = \frac{Ba_1 - Ab_1}{a_1b_2 - a_2b_1}$$

Make 260 pounds of high-protein trail mix and 280 pounds of high-carbohydrate trail mix for a maximum revenue of $1,880.

If in the unlikely chance that the largest revenue is obtained by using one of the numbers P, Q, or R and either of the associated values of X and Y are negative, eliminate that as a possible solution. Now look for the largest revenue from the remaining two numbers of P, Q, and R.

No one would deny that this is a lot of work, but this problem can appear in many different guises. Here's another variation of the same problem.

Practical Math Problem

4.5 Best Use of Two Ingredients in a Different Guise

A TV company makes dramas and reality shows, and has 120 actors available to appear in them. A drama costs $4,000,000 to produce and uses four actors, while a reality show costs $1,000,000 to produce and uses six actors. If a drama attracts 3,000,000 viewers and a reality show attracts 2,000,000 viewers, how should the company allocate its budget of $80,000,000 to attract the largest number of viewers? Each actor can only appear in one show.

Here the Products are the different types of shows, the Items are the budgets and the actors, and the Sales Price is the number of viewers attracted. The TV company should produce eighteen dramas and eight reality shows to attract a total audience of 70,000,000 viewers.

The Other Side of the Coin

The problems just presented involve maximizing revenue—although one problem measured the revenue in dollars; the other measured the revenue in viewers. There is another important aspect of business decisions: minimizing cost. We will encounter another version of these problems in Chapter 11.

Basic Business Percentages

We receive a great deal of information in percentage form, especially when it involves money. Wholesale prices rose 4% from the same quarter last year. The stock market fell 40% during its most recent downturn. Unfortunately, many people misunderstand percentages. Most people do understand what it means when we say wholesale prices rose 4% from the same quarter last year. There are two extremely important percentages that commonly occur in business: retail markup percentage and retail markdown percentage.

Retail Markup Percentage

This is also known by the term *margin percentage*.

Practical Math Problem

4.6 Retail Markup Percentage

The retail price of a digital camera is $90. The cost to the retailer is $55. What is the markup percentage?

➤ P = retail price

➤ C = cost to retailer

➤ M = markup percentage

$$M = \frac{100(P - C)}{P}$$

The retail markup percentage on the camera is 38.89%.

This problem illustrates the importance of knowing what the base is when computing a percentage. The markup percentage is the percentage of the retail price that the item has been marked up, not the percentage of the original cost by which the item has been marked up. In both computing and discussing percentages, it is always important to know what the base is on which the percentage is figured.

Retail Markdown Percentage

Sometimes prices are marked up; sometimes they must be marked down in order to move the merchandise.

Practical Math Problem

4.7 Markdown Percentage

A cell phone is marked down from $70 to $50. What is the markdown percentage?

➤ O = original price

➤ F = final price

➤ M = markdown percentage

$$M = \frac{100(O-F)}{O}$$

The cell phone has been marked down 28.57%.

It's important to note that the markdown percentage is computed on a base of the original price, whereas the markup percentage is computed on a base of the final price. This is the type of situation that can lead to misunderstandings and inaccurate computations.

Some Common Percentage Misunderstandings

Even so simple a misunderstanding as the units in which parts are being measured can result in major losses. In 1999, a Mars mission costing $125 million was lost because one of the engineering teams involved was measuring parts in the English system (feet and pounds), whereas another engineering team was using the metric system (meters and kilograms). There's a big difference: 3 meters is a lot longer than 3 feet, and 3 kilograms weighs a lot more than 3 pounds. If the bright engineers at NASA can make such obvious mistakes, it shouldn't be surprising that there are lots of opportunities lurking in combined percentage calculations to trap the unwary.

Combined Percentage from Two Successive Increases

Practical Math Problem

4.8 Two Successive Percentage Increases

The price of a stock appreciates by 10% 1 year and by 20% the next year. By what percentage has the price increased during the 2-year period?

➤ f = percentage of first increase

➤ s = percentage of second increase

➤ P = cumulative percentage increase

$$P = 100((1+.01f)(1+.01s) - 1)$$

The stock price increased by 32% over the two-year period.

This is not a tricky problem, but a lot of people get fooled and add percentages the same way they would add numbers. If the stock price went up $10 the first year and $20 the second year, it would obviously have increased by $30 over the two-year period.

While the formula suffices to answer the question, you can do this problem simply by assuming that the original stock price was 100. The original stock price really is 100—it just might not be 100 dollars. For instance, if the original stock price was $10, that's the same as 100 dimes. In the first year it goes up 10% of 100 to 110, but in the second year it goes up 20% of 110 to 132. It's the change of the base from 100 for the first increase to 110 for the second increase that is the source of confusion.

Combined Percentage from Two Successive Decreases

Frequently, when an item doesn't sell, it's marked down a percentage, and when it still doesn't sell, the price gets slashed yet again.

Practical Math Problem

4.9 Two Successive Percentage Reductions

A dress is discounted 30%, and then is discounted an additional 20%. What is the combined discount percentage from the original price?

➤ f = percentage of first discount

➤ s = percentage of second discount

➤ P = cumulative discount percentage

$$P = 100(1 - (1 - .01f)(1 - .01s))$$

The combined discount percentage of the two markdowns is 44%.

This problem can also be approached by the method described in the case of two successive price increases, by assuming that the original price is 100 and going from there.

Percentage Gain Needed to Recover a Loss

Another source of percentage misunderstandings occurs when an investment you made declines in value and you wish to compute the percentage that it needs to gain in order for you to get back to even.

Practical Math Problem

4.10 Percentage Gain Needed to Recoup Loss

An investment has declined in value by 20%. What percentage gain is needed in the investment in order for the investment to recover to its original value?

➤ L = percentage of loss in investment

➤ G = percentage gain needed to recover to its original value

$$G = \frac{100L}{100 - L}$$

The investment needs to increase in value by 25% to recover to the original price.

Once again, you can see this by assuming an original price of 100. A 20% decrease brings it down to 80, and from there it needs to increase by 20, which is 25% of 80, to get back to 100.

Percentage Loss Required to Get Back to Even After an Original Gain

This chapter closes on a happy note. You've made an investment that is doing nicely and want to know how much of a loss you can survive without going into the red.

Practical Math Problem

4.11 Percentage Loss to Fall Back to Even

An investment has increased in value by 20%. What percentage loss is required in order for the investment to decline to its original purchase price?

➤ G = percentage of gain in investment

➤ L = percentage loss required to fall to its original value

$$L = \frac{100G}{100 + G}$$

The investment can fall by 16.67% before you are in the red overall. You might notice that it seems like the percentage deck is stacked against you: if the investment loses 20%, you need a gain of 25% to get back to even, but if the investment gains 20%, it only has to fall by 16.67% to get you back to even.

CHAPTER 5

Borrowing Money and Payment Plans

In This Chapter

➤ Simple and compound interest

➤ Present value of a loan

➤ Buying a home or a car

Can I Skip This Chapter?

Money makes the world go round, but without the ability to borrow money and buy things on credit, the world would be going around a lot slower.

Most of the major purchases we make are done on credit, and it is safe to say that there would be a lot fewer houses being built and a lot fewer cars on the road if it were not possible to buy things on credit. Many of us even pay for minor purchases—a latte at Starbucks—by taking out our credit card. Credit cards are certainly a convenience, but they are also a temptation, seducing us into purchases we may not be able to afford because we know we can always put off paying until tomorrow by making minimum payments on our credit card.

Most of us really don't know very much about the mathematics of borrowing money. We know that lower rates are good, but that's about it. Even though this is not a math text in the technical sense of the word, whenever you use one of the formulas in this chapter, you'll learn something about the workings of credit. Properly used, credit is an engine of prosperity. Improperly used, it can be responsible for causing incalculable damage, as the once-in-a-century credit tsunami in the banking and housing industries in 2008.

If you are old enough to buy something on credit, this is an important chapter. If you are not yet old enough to buy something on credit but are old enough to read this book, this chapter is even more important because it's never too soon to learn the ins and outs of purchasing decisions.

When you borrow money, there are basically two ways you can pay it off. You can pay it off in a lump sum at the end of the term of the loan, or you can pay it off in a series of periodic payments. The first few topics of this chapter deal with simple interest and compound interest, which are ways of paying off the loan in a lump sum. The remainder of the chapter deals with periodic payments.

Simple Interest

The foundation of the credit structure is the idea of simple interest: if you borrow money, you should pay an amount proportional to both the amount of money that you have borrowed and the time that it takes you to pay it back.

Practical Math Problem

5.1 Simple Interest

How much interest will accrue to a loan of $2,000 earning 3% simple interest for a period of 4 years? What will be the total amount due to the lender at the end of the 4 years?

➤ P = amount of loan

➤ r = interest rate (expressed as a percentage)

➤ t = number of years

➤ I = accrued interest

➤ A = total amount due

$$I = .01Prt$$

$$A = P(1+.01rt)$$

The accrued interest at the end of the four years will be $240. The total amount to be paid to the lender at the end of that period is $2,240.

Simple interest presents the first opportunity in this book to see alternative forms of the same formula.

Alternative Forms

Alternative Forms

$$P = \frac{100I}{rt}$$

This form answers questions such as "How much money must be loaned at 2% interest for five years in order to earn $100 in interest at the end of that period?" The answer to this question is $1,000.

$$t = \frac{100I}{Pr}$$

This form answers questions such as "How long must the term of a loan of $4,000 at 5% interest be in order to earn $600 at the end of that period?" The answer to this question is three years.

Simple interest is easy to compute, but it's not in much use in the United States. Most loans that are made involve compound interest.

Compound Interest

Compound interest arose from a very natural realization that the person who loaned money for more than a year was being deprived of the opportunity to loan the interest that would have accrued at the end of a year.

Consider a loan of $1,000 made at 3% interest for two years. According to the simple interest formula, at the end of the first year, $30 interest would be due—but the borrower does not have to pay this back until the end of the second year. Admittedly, the $30 would only earn

90¢ in interest for the second year of the loan, but the borrower effectively has the use of this extra $30 without having to pay interest on it. Thus was born the idea of compound interest.

Interest can be compounded—the amount computed and then added to the amount of the loan—yearly, semi-annually, quarterly, monthly, or daily. In the latter case, there is some debate about how many compounding periods there are in a year: some loans specify 365 days in a year; others 360, a holdover from when it was easier to compute with 360 than with 365.

Practical Math Problem

5.2 Periodic Compound Interest

How much interest will accrue to a loan of $2,000 earning 3% interest compounded quarterly for a period of 5 years? What will be the total amount to be paid to the lender at the end of the 5 years?

➤ P = amount of loan

➤ r = annual interest rate (expressed as a percentage)

➤ t = number of years

➤ N = number of compounding periods per year

➤ I = accrued interest

➤ A = total amount due

$$I = P\left(\left(1 + \frac{.01r}{N}\right)^{Nt} - 1\right)$$

$$A = P\left(1 + \frac{.01r}{N}\right)^{Nt}$$

The accrued interest will be $322.37. The total amount to be paid to the lender is $2,322.37. In this case, the person who makes the loan earns an extra $22.37 by loaning it at 5% compounded quarterly than at 5% annual simple interest.

Alternative Forms

Alternative Forms

$$P = \frac{A}{(1 + \frac{.01r}{N})^{Nt}}$$

This formula answers questions such as "How much must be deposited in order to have $5,000 in four years at 2% compounded semi-annually?" The answer to this question is $4,617.42.

$$t = \frac{\ln(\frac{A}{P})}{N\ln(1 + \frac{.01r}{N})}$$

This formula answers questions such as "How much time is needed for $5,000 to become $8,000 at 3% compounded quarterly?" The answer to this question is 15.73 years.

Banks usually have early withdrawal penalties, so you'll have to leave it in for 15.75 years and receive a little more than $8,000.

The formula for P in the Alternative Forms sidebar is especially important. The amount $4,617.42 is known as the present value of $5,000 in three years at 2% compounded semi-annually. When an individual or an institution has a number of loans outstanding for different periods and at different rates, it computes the value of those loans by adding up the present values of the loans.

Converting Compounding Rates with Different Frequencies

When trying to borrow money, the borrower is often confronted with different rates at different frequencies (the number of compounding periods in a year). How can you tell which is the better deal? A relatively straightforward way to do this is to compute the present value of $1,000 for each of the different options with which you are presented. If there are no other considerations, you should choose the one that has the highest present value because the difference between the $1,000 at the end of the year and the present value now represents the amount of interest that you are paying. The smaller that difference, the less interest you will have to pay.

Practical Math Problem

5.3 Different Compounding Frequencies

What interest rate compounded semi-annually is equivalent to a 2.2% rate compounded monthly?

➤ r = original interest rate as percent

➤ n = annual number of compounding periods for original rate

➤ R = new interest rate as percent

➤ N = annual number of compounding periods for new rate

$$R = 100N\left(\left(1 + \frac{.01r}{n}\right)^{\frac{n}{N}} - 1\right)$$

It is equivalent to 2.21% compounded semi-annually. *Equivalence* means that money invested in two different ways will result in the same amount at the end of the year.

Continuous Compounding

Let's take a look at what happens if $1,000 is compounded at 4% for a year at different frequencies:

➤ Annually: $1,040.00

➤ Semi-annually: $1,040.40

➤ Quarterly: $1,040.60

➤ Monthly: $1,040.74

➤ Daily (365 days): $1:040.81

The more frequently we compound, the more interest we make. But it seems as if there is a limit to what we can make, even if we were to compound every nanosecond. This type of compounding is known as continuous (or instantaneous) compounding, and it is important not only in finance, but it is also appears in natural processes such as unchecked growth and radioactive decay (which will be discussed later).

Practical Math Problem

5.4 Continuous Compounding

How much interest will accrue to a loan of $2,000 earning 2.5% interest compounded continuously for a period of 8 years? What will be the total amount to be paid to the lender at the end of the 8 years?

➤ P = amount of loan

➤ r = annual interest rate (expressed as a percentage)

➤ t = number of years

➤ I = accrued interest

➤ A = total amount due

$$I = P(e^{.01rt} - 1)$$

$$A = Pe^{.01rt}$$

The accrued interest will be $442.81. The total amount to be paid to the lender is $2,442.81.

The formulas above include the factor .01, to convert the percentage rate r to the decimal rate .01r. Most people encounter rates as percentages. When this formula is presented in textbooks, the decimal rate is almost always the one that is used; percentages are useful for commerce but are an artifact of an ancient method of collecting taxes based on the number 100. When the decimal rate r is used, the last formula is A = Pe^{rt} and is often referred to as the PERT formula (just so you know should you encounter it).

Alternative Forms

Alternative Forms

$$P = Ae^{-0.01rt}$$

This formula answers questions such as "How much must be deposited in order to have $5,000 in four years at 2% compounded continuously?" The answer to this question is $4,615.58. As before, this is called the present value of $5,000 in four years at 2% compounded continuously.

$$t = \frac{100 \ln(\frac{A}{P})}{r}$$

This formula answers questions such as "How much time is needed for $5,000 to become $7,500 at 3% compounded continuously?" The answer to this question is about 13.52 years.

Buying Things by Making Periodic Payments

Making a series of periodic payments to purchase something is a key feature of a credit economy. We mentioned earlier that few cars and almost no houses could be purchased were it not for the willingness of those with the financial resources to loan money for these purchases.

It is very likely that you have already purchased something on such a credit plan. Most credit plan purchases are done with monthly payments: small purchases are generally paid for within a year or so, cars generally over a four- or five-year period, and houses typically over a thirty-year period. In a large number of such purchases, a down payment of a percentage of the purchase is required on the signing of the contract, with the first monthly payment due a month later.

Although a certain amount of credit worthiness must be shown by someone who wants to borrow money from a financial institution, the threshold is higher for larger purchases such as houses, and the amount of money that can be loaned depends on what is called the loan-to-value ratio (which was briefly discussed in Chapter 1).

Practical Math Problem

5.5 Loan-to-Value Ratio

A bank approves a loan of $450,000 at a loan-to-value ratio of 90%. What is the maximum appraised value of a property that can be purchased with this loan?

➤ L = loan amount

➤ P = loan-to-value percentage

➤ A = maximum appraised value of property

$$A = \frac{100L}{P}$$

The maximum appraised value of the house is $500,000. Of course, this assumes that the borrower can make the 10% down payment that this would require.

Now let's look at one of the most common types of periodic payments: the five-year auto purchase plan.

Practical Math Problem

5.6 Periodic Payments

What is the monthly payment for a car that costs $12,000 if a 10% down payment is made and the balance is to be paid in monthly installments over a 5-year period with money borrowed at 6%, and the first payment is due a month after the car is purchased?

➤ C = cost of car

➤ d = percentage of down payment

➤ r = annual loan percentage

➤ n = number of payments per year

➤ N = number of years for loan

➤ P = periodic payment

First calculate the periodic interest rate, or the per-period interest rate.

$$i = \frac{.01r}{n}$$

Then compute the periodic payment.

$$P = \frac{(1-.01d)Ci}{(1 - \frac{1}{(1+i)^{nN}})}$$

The monthly payment necessary to purchase the car is $208.79.

This formula can also be used to find out how much can be withdrawn periodically from a fixed sum (such as an inheritance) if the total amount is to be withdrawn after a predetermined time.

Balance Remaining on a Loan: Amortization

Let's say you have purchased the car in the previous example. Because you made a down payment of 10% ($1,200 in this example), you currently have a balance of $10,800 remaining on the loan. The per-period interest rate for 6% paid monthly is 0.5%, or one-half of 1%. If you borrow $10,800 and pay 0.5% interest, the amount of interest is $54. Your actual payment is $208.79; $54 goes to pay off the interest, and the remaining $154.79 pays down the balance of the loan. This $154.79 is subtracted from the $10,800, leaving $10,645.21 as the balance remaining. When you make your next payment, the interest of 0.5% is charged on the $10,645.21 and amounts to $53.23. Of your second payment of $208.79, $53.23 goes to interest, and the remaining $155.56 pays down the balance of the loan, reducing it from $10,645.21 to $10,489.65. This process is known as amortization.

Practical Math Problem

5.7 Balance Remaining on a Loan

What is the balance remaining on a loan of $20,000 if the money was borrowed at an annual rate of 5% and 24 monthly payments of $500 have been made?

Be prepared; this results in a very lengthy formula.

A = amount of loan

r = annual loan percentage

n = number of payments per year

N = number of payments that have already been made

P = periodic payment amount

B = loan balance

Again, the first step is to calculate the periodic interest rate.

$$i = \frac{.01r}{n}$$

Then calculate the remaining balance.

$$B = A(1 + i)^N - P\frac{((1 + i)^N - 1)}{i}$$

The balance remaining on the loan is $9,505.87.

When a loan is refinanced, it is the balance remaining on the loan that becomes the new loan. What typically happens is that a credit company puts aside sufficient cash to make the payments on the balance remaining on the loan according to the existing payment schedule. It then offers you a new loan with new terms such as a different interest rate and a different amount of time for the loan to be fully paid off.

Watch for Deceptive Re-Fi Claims!

If the term of a loan is reduced, say from fifteen years to ten years, what typically happens is that the individual payments—the checks you write out monthly—are increased. However, the total of these payments is less than the total of the payments on the original loan. This is the source of claims such as, "You can save hundreds of thousands of dollars by refinancing your loan." In one sense this is true—the total amount of the checks you write may be hundreds of thousands of dollars less than the ones you would have written out under the terms of the original loan.

Here's the tricky part: you should compute what the new loan does for you not on the basis of the total of the checks you write out, but on the basis of the total of the present values of all the checks you would write out. It's the total of the present values that counts, not the total of the checks you write out.

Funding a Perpetuity

Here's a way to see the reasoning behind the warning in the section above. A perpetuity is a fund that dishes out a fixed amount of cash every month (or every year)—forever! That's the *perpetual* in perpetuity: the fund promises to pay the money out until the end of time (or the universe, whichever comes first).

Practical Math Problem

5.8 Funding a Perpetuity

How much is needed to fund a perpetuity that will pay $10,000 annually if the money is invested at 2.5% compounded annually?

➤ P = annual payout

➤ I = investment return in percent

➤ A = amount needed to fund the perpetuity

$$A = \frac{100P}{I}$$

You need $400,000 to fund the perpetuity.

How does this relate to the warning in the section above? The total of the checks written out by the perpetuity is, at least in theory, infinite, but given the choice of having $500,000 in the present when interest rates are 2.5% or having $10,000 forever, it's better to take the $500,000.

Incidentally, the same advice generally applies if you win the lottery. Take the lump sum payment rather than the annual payouts.

How Long Will an Inheritance Last?

Many of us will inherit money, and a common plan is to invest it in a safe manner and withdraw a fixed amount every year. The obvious question is how long the inheritance will last.

This problem is related to the problem of financing a car or a house. In those cases, the term of the loan is predetermined, and doing the math consists of finding how much each payment should be. In an inheritance situation, the amount of each payment is fixed, and the problem is to determine how long the inheritance will last.

Practical Math Problem

5.9 How Long Will an Inheritance Last?

How long will an inheritance of $100,000 last if it is invested at 3% and $7,500 is withdrawn as soon as the inheritance is received and at the beginning of every year thereafter?

➤ I = amount of inheritance

➤ r = annual interest rate

➤ P = periodic withdrawal

➤ N = number of years before inheritance is depleted

$$N = \frac{\ln(\frac{P(1+r)}{P(1+r) - Ir})}{\ln(1+r)}$$

The inheritance will last 16.63 years, which means that the last payment will be somewhat less than the $7,500. It's a lengthy calculation, but it's a calculation you will want to make to be sure that you know how long your inheritance will last.

CHAPTER 6

 # Stocks, Mutual Funds, Bonds, and Options

In This Chapter

➤ Risk and reward

➤ Return on investment annualized and real rate of return

➤ Different investment vehicles

Can I Skip this Chapter?

This chapter is useful for anyone who invests in the market—which means practically everyone nowadays. It is especially useful if you make your own investment decisions, but everyone else will find reading it worthwhile—especially the section on dollar-cost averaging.

History of the Stock Market

The stock market is another one of the engines of capitalism that have enabled our economy to be productive. The first stock markets were founded with a relatively simple idea in mind: Large projects were beyond the resources of a single investor or even a group of investors. It was desirable to have a supply of capital available for the undertaking of such projects (Notably sea explorations of the seventeenth century, which cost a lot of money, were notoriously risky, but were highly profitable if successful.). These projects were difficult to fund by tapping a few investors, even well-heeled ones. However, by opening the door to the public, a large number of individual contributions would enable these ventures to be funded. Maybe this is how the term *venture capitalism* first arose.

This was a remarkably successful idea for several reasons. Not only did it enable the ventures to obtain enough capital, it allowed individuals to participate in them. A single investor could expose himself to whatever amount of risk he deemed appropriate. If the ship sank, there went the investment, but a ship that returned laden with treasure could return multiples of the original investment.

Today's capital markets extend far beyond the reaches of the first stock markets. Entire libraries are devoted to each of the topics that will be mentioned in this chapter, and there are numerous topics, such as the commodity markets, that this chapter doesn't even mention. Nonetheless, this chapter is a good place to get started. Each topic will have some introductory material for those who may not be familiar with the basic idea, and then some of the practical math associated with that topic will be presented.

Risk and Reward

One thing hasn't changed since those first early sea explorations were funded. If you make any sort of investment, you are taking a risk in the hope of obtaining a reward.

Limited and Unlimited Risk

Limited risk means that the amount of money you could lose is limited to the amount that you invest. When you buy shares of stock, the risk is limited. If the company in which you bought stock goes bankrupt so that the value of the stock is zero dollars per share, you've lost what you've invested. The company's creditors are not going to come knocking on your door seeking to repossess your car or house in order to satisfy the company's debts.

As long as you buy stocks, mutual funds, bonds, or options, your risk is limited to the money you have invested. You can risk a little more than the amount you actually put up to make the transaction; this is known as *buying on margin*, which will be explained later.

You can risk a lot more if you engage in certain transactions such as selling naked options. Even though the risk is still theoretically limited, you are sometimes risking 10 or even 100 times what you stand to make, so the term *unlimited risk* is used to describe this.

Unless you are a sophisticated investment professional, you should only take limited risk with a fixed fraction of your available capital. Treat investing like a trip to Vegas: you hope to win, but if you lose you'll still be able to pay the rent. Leave unlimited risk investing to the pros.

Stocks and Mutual Funds

If you listen to the stock market report on the TV, or check the market online or in the paper, every day there are reports of companies that either fail completely or take a severe hit to their stock price. A movie company with a string of flops or a retail store with a bad Christmas season is almost certain to see its price decline—and sometimes it will decline a lot. Stocks you buy almost always seem to go down more than you expect and go up less than you hope. Nonetheless, investing in stocks is actually a good idea overall, or at least it has been historically. The overall market—the Dow Jones average, for example—has outperformed inflation significantly over the long run, although there are periods when it takes a bad hit.

The easiest way to reduce your risk is through diversification. If you buy a number of different stocks, your overall risk is less—but your overall reward is lower, too. An individual stock can triple in price or lose all its value during a single year, but the Dow Jones average rarely gains or loses more than 20% annually.

A mutual fund is a collection of different stocks. By buying shares in a mutual fund, you are limiting both your risk and your reward because the performance of a group of stocks is always less extreme than the performance of the most volatile stocks in the group.

There are all sorts of mutual funds. There are mutual funds that are linked by a common denominator, such as biotech stocks or Japanese stocks. There are mutual funds run by investment professionals; their past performance (which, as the disclaimers say, is no guarantee of future performance) is a matter of record. There are index funds, which buy every stock in an index such as the Dow Jones average in the same proportion as the stock appears in the index. Such funds mirror the index; they do exactly as well, or as badly, as the index itself.

Keeping Score: The Real Rate of Return and ROIA

You bought a stock or a mutual fund. Later, you sell it back at a profit. That's good—but how did you really do?

What is needed is some sort of a benchmark to indicate how your investment actually performed. You can compare it with any number of things. One possibility is to compare your investment with the market as a whole. However, the real question is, are you better off for having made the investment, and, if so, how much better off are you? That question is answered by a measure called the real rate of return.

Practical Math Problem

6.1 Real Rate of Return

Between October 2009 and October 2010, the rate of inflation was 1.18%. During that period, the price of gold increased by 17.39%. What was the real rate of return on gold during that period?

➤ r = annual percentage rate of actual commodity

➤ i = annual percentage rate of inflation

➤ R = percentage of real rate of return

$$R = 100\left(\frac{1+.01r}{1+.01i} - 1\right)$$

The real rate of return of gold during this period is 16.02%.

The real rate of return uses inflation as a benchmark, tracking how much your investment improved or fell behind inflation. If you did better than inflation, your real rate of return is positive; if you did worse than inflation, your real rate of return is negative. If your investment made 3% but the rate of inflation is 4%, inflation outperformed your investment. Even though you made money, at the end of the year your purchasing power is less.

The real rate of return is a powerful analytical tool, but it is generally used on an annual basis because it is felt that a full year gives a more accurate indication of inflation than a shorter period. Nonetheless, many investments in the stock market are closed out by selling what was originally bought in periods much shorter than a year. The return on investment annualized, colloquially known as ROIA, is designed to see how such an investment does on a yearly basis.

Practical Math Problem

6.2 Return on Investment Annualized

A stock is purchased for $101 and sold 60 days later for a price of $104.75. The stock also paid a dividend of 25¢ during that period. The commissions for buying and selling amounted to 48¢ per share. What is the ROIA for this transaction?

> ➤ P = purchase price of stock
>
> ➤ S = sale price of stock
>
> ➤ C = commission costs
>
> ➤ D = dividends collected (if any)
>
> ➤ N = number of days to complete transaction
>
> ➤ R = return on investment annualized (expressed as a percentage)

$$R = 100((\frac{S + D - C}{P})^{\frac{365}{N}} - 1)$$

The ROIA for this transaction is 23.17%.

In general, stocks are not purchased one share at a time; a standard purchase volume for stocks is one hundred shares, and commissions are charged on this basis. *Odd lots*, which are purchases of less than one hundred shares, have a different commission schedule. At any rate, the commission in the above problem should be the commission per share.

Practical Tools for Evaluating Stocks

This section comes with the usual disclaimer that the formulas in this section by themselves are insufficient to determine whether a particular stock will go up or down. Nonetheless, they are tools used by almost everyone who attempts to analyze stocks on the basis of whether it is fairly priced with respect to the health of the underlying company.

This is also our first look at a different type of important and practical mathematical tool: the ratio. A ratio is a quotient of two quantities constructed for one of several purposes. Some ratios give insight into the behavior of certain parameters; others help to compare one entity with another.

Practical Math Problem

6.3 Dividend Yield Ratio

A stock that is currently trading at $80 per share pays quarterly dividends of 48¢. What is the dividend yield ratio?

➤ P = price of stock

➤ D = yearly total for dividends

➤ Y = dividend yield ratio

$$Y = \frac{D}{P}$$

The stock has a dividend yield ratio of 0.024. Translated to a percentage, this means that the stock pays 2.4% of its share price in dividends. If inflation stays under 2%, as long as this stock does not decline in price and continues to pay the dividend (bad news: stocks sometimes cut their dividend or don't pay it), the stock shows a positive real rate of return.

The dividend yield ratio helps to judge directly whether a stock is a good investment. Stocks with high dividend yield ratios are obviously attractive candidates for investment.

Practical Math Problem

6.4 Earnings per Share

A stock has 40,000,000 outstanding common shares (*outstanding* is a term meaning the number of shares issued by the company). The company has annual earnings of $2,500,000. What are this stock's earnings per share (abbreviated EPS)?

➤ N = number of outstanding common shares

➤ I = annual net income

➤ E = earnings per share

$$E = \frac{I}{N}$$

The stock has an EPS of 6¢ per share (these are usually rounded off to the nearest cent).

This particular ratio is sometimes called the *undiluted* EPS, which is contrasted with the *diluted* EPS in which shares available from convertible bonds, vested options, and warrants are also included in the denominator.

EPS values are generally compared within industries rather than between industries, and can help to judge whether one company's stock is a better value than another's within the same industry.

Practical Math Problem

6.5 Price/Earnings Ratio

A stock is currently trading at $40 per share and has earnings of $2.40 per share. What is its price/earnings ratio (abbreviated P/E)?

➤ P = current price of stock

➤ E = earnings per share

➤ R = price/earnings ratio of the stock

$$R = \frac{P}{E}$$

The stock has a price/earnings ratio of 16.67.

The P/E ratio of a stock is considered so important that it is included in the first page of practically every report one can find on the stock. As a generality, a low P/E may be an indicator that the company's shares are undervalued, whereas a high P/E might indicate just the reverse.

However, what you would probably like is a strategy that will help you show a profit in the stock market. Here is a time-tested one, with the usual legal demurrer about past performances, etc.

Dollar-Cost Averaging

Dollar-cost averaging is a long-term strategy. It involves investing a constant amount monthly in the same equity. For the purposes of argument, let's assume that equity is a Dow Jones average fund, and you are going to invest $200 per month. When the Dow Jones goes up, your shares are worth more; when it goes down, your monthly investment buys more shares. Assuming (here's where the disclaimer kicks in) that the Dow Jones beats inflation over the time you pursue this strategy, you will show a positive real rate of return.

Many people who have taken this approach from the time they were young looked at their portfolio when they hit their fifties and realized they could comfortably retire.

Buying on Margin and Margin Calls

Buying on margin is borrowing money from your broker to purchase stock. During the Roaring Twenties, it was possible to purchase stocks on 10% margin. That means that if you wanted to buy a stock trading at $50, you merely had to give your broker 10% of $50 for each share of stock you wanted to purchase on margin.

If the stock went up to $55—a 10% rise in price—you could sell the stock and make $55, $50 less commissions, nearly doubling your money! However, the dark side of the picture was that if the stock went down to $45, you lost all your money. And if it went down below $45 you were liable for more money than you had actually invested. If the stock suddenly became worthless, you would lose $50 on a $5 investment. This might be called a limited risk, but only in the technical sense of the word.

Now look at it from the standpoint of the broker. As soon as the stock starts to sink below $50, the broker worries that it might go even lower and you might not be able to pay. So the broker sends out a margin call, essentially a demand that you come up with more money— right now! If you can't, the broker will sell the stock to limit your loss and the broker's risk of having loaned you the money.

The stock price at which the broker sends you the dreaded margin call is known as the margin call trigger price.

Practical Math Problem

6.6 Margin Call Trigger Price

A stock selling at $80 is purchased for an initial payment of 70%. If the maintenance margin is 55%, at what price will a margin call be triggered?

> ➤ P = purchase price of stock
> ➤ I = initial margin percentage
> ➤ M = maintenance margin percentage
> ➤ C = margin call trigger price

$$C = \frac{100 - I}{100 - M} P$$

A margin call is triggered when the price of the stock drops to $53.33.

As a result of the market crash of 1929, the Securities and Exchange Commission changed the rules on margin buying. You are now limited to buying on 50% margin, but, as in the previous example, you don't have to buy on 50% margin; you can put up a higher percentage of the initial price.

Incidentally, the broker will charge you interest on the money he loans you. Make sure this is included when you compute your ROIA and real rate of return.

Bonds

A bond is a promise to pay a fixed amount—usually $1,000—on a certain day. Bonds are issued by various organizations that seek financing; essentially, they are seeking loans from the bond purchaser. The bond yield is the annualized percentage return the holder of the bond will receive if the bond is held until maturity, the day on which the fixed amount is paid to the bond holder.

Most bonds also pay an annual amount, analogous to the dividend a stock pays. This annual amount is called the bond's coupon. If you've ever heard the expression "clipping coupons," it doesn't refer to clipping "40 cents off on peanut butter" coupons from the newspaper. Bonds were originally issued with detachable coupons. To claim your annual payment, you submitted a coupon.

Because a bond promises to pay $1,000 at maturity and also issues a coupon, bonds are almost invariably bought at a lower price than their face value. How low the price is depends upon a number of factors, including the reputability of the organization that issued the bond. In order to protect the unwary consumer against bonds issued by less credit-worthy organizations, there are various bond-rating agencies. The best known of these agencies are Standard & Poor's and Moody's. The rating scales of the agencies differ, but they're like school grades, except that a grade of C is basically not passing. Bonds with ratings that have the letter C (or worse) in them are known as junk bonds.

As with all investments, a bond's reward and risk are related. The safer the bond, the lower the yield. The riskier the bond, the higher the yield.

Practical Math Problem

6.7 Yield to Maturity

A bond has 5 years until it matures at a face value of $1,000. It pays an annual coupon of $40 and is currently priced at $950. What is the yield to maturity?

➤ C = annual coupon

➤ F = face value

➤ P = price

➤ N = number of years until maturity

➤ Y = yield to maturity (in %)

$$Y = \frac{200(NC + F - P)}{N(F + P)}$$

The yield to maturity is 5.13%.

Bonds can be issued by corporations and by federal, state, and local agencies. In order to promote the latter, the interest on some bonds is exempt from taxes. Normally, such bonds will have a lower yield than a comparable bond (in the ratings sense) issued by a corporation.

A reasonable question for an investor to ask is whether the yield from a bond that is subject to taxes is a more attractive purchase than the yield from a tax-exempt (a.k.a. tax-free) bond. The following formula addresses this question.

Practical Math Problem

6.8 Equivalent Taxable Yield

A bond has a tax-free yield of 1.45%. What is the equivalent taxable yield for a person with a marginal tax rate of 28% (a.k.a. an income tax bracket of 28%)?

➤ F = tax-free yield (in percent)

➤ T = equivalent taxable yield (in percent)

➤ R = marginal tax rate (in percent)

$$T = \frac{100F}{100 - R}$$

The equivalent taxable yield is 2.01%.

Alternative Forms

Alternative Forms

$$F = T(1 - .01R)$$

This form answers questions such as "What tax-free yield is equivalent to a taxable bond paying 2.1% for someone in the 33% tax bracket?" The answer to this question is 1.41%.

Options: Calls and Puts

An option is an example of a derivative; it depends on the price of something else. Specifically, a call option confers the right, but not the obligation, to purchase a stock on or before a certain date at a given price. A put option similarly confers the right, but not the obligation, to sell a stock on or before a certain date at a given price. It sounds a little complicated—let's take a look.

Call Options

As of this writing (February 2011), IBM is trading at a little more than $164 a share. You can buy an IBM April 170 for $191, which allows you to buy 100 shares of IBM any time between now and April 15th, when the option expires, for $170 a share. The share price you're willing to pay, in this case $170, is known as the strike price of the option.

Why would you make this purchase? Let's say that on April 15th, IBM is trading for $180 a share, a gain of slightly less than 10% from where it is now. You could exercise your option, buy 100 shares of IBM for $170 a share, and turn around and sell them in the market for $180 a share. You would make a profit of $10 on each of your 100 shares, or $1,000. Subtract the $191 you paid for the option and you've made a little more than $800 (less commissions)—you've quadrupled your money.

Unlike trading on margin, your risk is limited to the money you spend to buy the option. The options market is highly liquid; every day the price of your option will change, and you don't have to wait until April 15th to sell your option. The analysis of the value of stock options is a sophisticated example of high-powered math, and we won't go into it here.

Put Options

A put option is similar to a call option, except it gives you the right to sell 100 shares of stock at a given price on or before a certain day. You can buy an IBM April 155 put for $137, which gives you the right to sell 100 shares of stock at $155 on or before April 15th. If IBM does badly and goes down to $150, you can exercise your option, sell 100 shares at $155 and buy 100 shares at $100, netting $5 a share, or $500. Subtract the $137 you paid to buy the put and your profit is $363 (less commissions)—you've nearly tripled your money.

Buying either call or put options is risky; even though you can make multiples of your money, you can lose all your money. However, there is an attractive strategy if you already own stock that goes by the name of writing covered calls.

Suppose you own 100 shares of IBM at this moment. You could sell the April 170 call discussed above for $191. If IBM never goes above $170, that's an extra $191 in your pocket.

On the other hand, if the stock closes above $170 on April 15th, the owner of the option will exercise it, pay you $170 per share for the stock, and take your stock. By selling this option, you limit your potential for a huge gain, as you cannot make any more money if the stock closes above $170, but you have an extra $191 if it doesn't.

Practical Math Problem

6.9 ROIA on Covered Calls

IBM is currently trading at $164. You sell an IBM April 170 call for $1.91 a share; the actual price is $191 as the option is for 100 shares of stock. There are 60 days until expiration. What is the annualized return on the option if you still own the shares at expiration?

➤ C = current price of stock

➤ P = option price per share

➤ N = number of days to expiration

➤ S = strike

➤ A = annualized return on investment as percentage

$$A = 100\left(\left(\frac{S}{C-P}\right)^{365/N} - 1\right)$$

The ROIA on writing covered calls is about 33.6%, so you can see why writing covered calls is an attractive strategy.

The calls are described as covered because you own the stock and can deliver it if someone exercises the option you sold. If you do not own the stock, you have sold what is known as a naked call, because you are exposed to theoretically unlimited risk. Here's where the injunction never to take unlimited risk is so important. Suppose that you have sold a naked April 170 call, IBM makes some remarkable discovery, and overnight the price jumps from $165 to $210 (these things happen). The call that you sold for $191 is now worth $4,000— and you must buy it back! Old saying in the market: He who sells what isn't his'n must buy it back or go to prison. Fortunately, brokerages generally do not allow you to sell naked calls unless you are a market professional— if you can't come up with the money, the brokerage is on the hook for it.

CHAPTER 7

Taxes and Other Governmental Math

In This Chapter

➤ The effect of taxes on individual and business decisions

➤ Depreciation

➤ When should you take Social Security?

Can I Skip This Chapter?

This chapter should be read by anyone who pays taxes. It is also extremely useful if you own or manage a business.

Tax-Deductible Items

Benjamin Franklin, one of the great figures in American history, uttered one of the most famous quotes about government when he said, "In the world nothing can be said to be certain except death and taxes." Medicine has made incredible inroads against death, but we still have yet to make any noticeable progress against taxes.

We saw in the last chapter the effect that taxes have when examining the purchase of tax-exempt bonds. Many aspects of investing are greatly influenced by taxes. One of the most important of these is the purchase of a house. In order to encourage home ownership, generally considered to be a critical component of the American dream, the portion of the monthly payments that goes to interest is tax deductible.

You probably know what *tax deductible* means, but in case you don't: if a purchase is tax deductible, the cost of that purchase may be deducted from your income when you are figuring the taxes you owe. Many expenses are tax deductible; the simplest one being contributions to charity. If you contribute $100 to a recognized charity such as the Red Cross, that $100 is deducted from your income when the time comes to compute your taxes.

Let's say you are in the 28% tax bracket and contribute $100 to the Red Cross. Had this contribution not been tax deductible, your reported income would have been $100 higher, and you would have not only paid the $100 that you contributed to the Red Cross, you would also have paid 28% of the extra $100 in your reported income to Uncle Sam. That's $28.

Practical Math Problem

7.1 Tax-Deductible Contributions

What is the true cost of a $100 contribution to a tax-deductible charity for an individual in the marginal 28% tax bracket?

➤ C = dollar amount of tax-deductible contribution

➤ R = tax bracket as percent

➤ T = true cost of contribution

$$T = (1 - .01R)C$$

The true cost of the Red Cross contribution discussed above is $72. You can think of it this way: had the contribution not been tax deductible, the government would have extracted another $28 from you. However, since it was, they returned the $28 to you, which you subtract from the actual cost of the contribution to compute the true cost.

Numerous items, in addition to contributions to qualifying charities, are tax deductible. A tax specialist should be consulted on these matters, although typing *tax-deductible items* into your browser will locate numerous articles on the subject.

Now let's look at the effect of the government's decision that the interest paid on qualifying purchases (such as a house) has on your bottom line.

Practical Math Problem

7.2 Net Monthly Payments

What is the net monthly payment after 120 monthly payments of $2,147.29 have been made on a 30-year loan of $400,000 at 5% for a person who is in the marginal 28% tax bracket? The net monthly payment is the payment less the amount saved due to the deductibility of interest.

➤ A = amount of loan

➤ r = annual loan percentage

➤ n – number of payments per year

➤ N = number of payments that have already been made

➤ P = periodic payment amount

➤ t = percentage of marginal tax bracket

➤ M = net monthly payment

First calculate the periodic interest rate.

$$i = \frac{.01r}{n}$$

We can now calculate the net monthly payment.

$$M = P - t\frac{Ai(1+i)^N - P((1+i)^N - 1))}{100}$$

The net monthly payment is $1,797.69.

In 2011, the 25% marginal tax bracket for single individuals went from about $34,000 to $83,000, and for a married couple filing jointly went from $69,000 to $139,000. A lot of people fall into this category. Let's see how tax deductibility affects them by using another tool of practical mathematics—the chart.

True Monthly Net Payment Percentage

This chart assumes that a thirty-year loan has been made for someone in the 25% marginal tax bracket.

		Percentage	Rate	
Years	4	5	6	7
1	83	81	79	78
5	84	82	80	79
10	86	84	82	81
15	88	87	85	83
20	91	90	88	87
25	95	94	93	92
30	99	99	99	99

This chart tells you that, in the fifth year of a loan made at 6%, the net payment is only 80% of the actual payment. An important thing to note about this chart is that for the first ten years of the loan, this percentage hardly increases at all, which means that for the first ten years of the loan, the borrower can count on the net monthly payment staying roughly the same. This is because most of the money used for payments during the first ten years of the loan goes to interest, rather than to reducing the balance of the loan.

Quarterly Estimated Tax Payments

The IRS not only wants your money, it wants it before you've made it! We pay estimated taxes in advance. If you are on a payroll, the estimated taxes are automatically deducted. This prevents you from earning interest on it and lets the government earn the interest. If you are self-employed, you have to pay estimated taxes in advance or incur a large penalty if you don't.

However, if you are self-employed, the government gives you a choice of two ways of paying your taxes. You can make a lump sum payment at the end of the year and incur a small penalty, or you can pay quarterly estimated taxes by paying roughly the same amount at the end of each quarter. There's a right way to go about this.

Practical Math Problem

7.3 Quarterly Tax Payments

In one scenario, you pay quarterly estimated payments of $6,000. In the other scenario, you pay a lump sum at the end of the year at a penalty of $100. The money you would have used to pay quarterly taxes could instead be put in a CD that earns 2% annually. Should you make the quarterly payments or put the money in a CD and pay a lump sum at the end of the year?

➤ Q = estimated quarterly payment

➤ P = penalty for end-of-year payment

➤ R = annual CD rate expressed as a percentage

$$X = \frac{200P}{3R}$$

If Q is greater than X, do not make the quarterly payments, otherwise do so.

In problem 7.3, Q is $6,000 and X is about $3,300, so you should pay the lump sum at the end of the year and put the quarterly payments in a CD.

This is an approximate answer, assuming that you collect simple interest rather than compound interest; over this short period of time, the difference won't matter much. It's a lot easier to calculate as well.

Depreciation

The value of an asset generally declines as it gets older. Personal assets such as cars lose their value; business assets such as heavy machinery are less effective and require more maintenance as they age. There are tax deductions available for depreciation of an asset, but this is a subject far too complex for a book like this to go into in any detail. Suffice it to say for our purposes that there are some general parameters that are common to depreciation situations:

➤ The original cost of the asset

➤ The salvage value of the asset

➤ The expected life of the asset

The expected life of the asset generally depends upon what category the asset belongs to. Office furniture has a different expected life than heavy equipment, and the IRS has mandated the expected lives for the various categories. The cost is known to you, and you can make a reasonable estimate of the salvage value.

You also have a choice of the bookkeeping method that you can use to depreciate the asset. These methods give different depreciation schedules for the expected life of the asset, which result in different tax consequences during those different years. Probably a wise thing to do is to make schedules for all of the depreciation formulas and look at the tax consequences of each. Some individuals or businesses would prefer to accept the greater tax burden in the near term, whereas others would prefer a different arrangement.

Straight-Line Depreciation

Straight-line depreciation is the easiest to compute and to understand: it depreciates the asset by the same amount each year.

Practical Math Problem

7.4 Straight-Line Depreciation

What is the annual straight-line depreciation on a computer purchased for $600 with an expected life of 4 years and a salvage value of $80? What is the book value at the end of the 3rd year?

> ➤ C = cost of equipment
> ➤ N = number of years of expected life
> ➤ S = salvage value
> ➤ D = annual depreciation
> ➤ n = number of years of depreciation
> ➤ B = book value after n years

$$D = \frac{C - S}{N}$$

$$B = C - \frac{n(C - S)}{N}$$

The computer depreciates by $130 every year. The book value, which is the value that the computer has for accounting purposes, at the end of the third year of depreciation is $210.

Double Declining Balance Depreciation

Double declining balance depreciation accelerates the amount of depreciation in the early years of the expected life of the asset. An asset with a four-year life span, an initial cost of $1,000, and no salvage value will have $250 of depreciation expense during the first three years. In contrast, using the double declining balance method, the depreciation expense during the first year is $500, during the second year $250, and during the third year $125.

Practical Math Problem

7.5 Double Declining Balance Depreciation

A truck is purchased for $20,000. It has an expected life of 10 years and a salvage value of $3,000. What is the book value at the end of the 4th year, and what is the depreciation for the 4th year, and as computed by the double declining balance method?

➤ C = cost of equipment

➤ N = number of years of expected life

➤ S = salvage value

➤ n = number of years of depreciation

➤ D = depreciation for year n

➤ B = book value after n years

$$B = \max\left(C\left(1 - \frac{2}{N}\right)^n, S\right)$$

$$D = \max\left(C\left(1 - \frac{2}{N}\right)^n, S\right) - \max\left(C\left(1 - \frac{2}{N}\right)^{n-1}, S\right)$$

For most of the lifetime of the equipment, the book value will not fall below the salvage value. In this case, the formulas are a little simpler:

$$B = C\left(1 - \frac{2}{N}\right)^n$$

$$D = C\left(1 - \frac{2}{N}\right)^{n-1}\left(\frac{2}{N}\right)$$

The book value at the end of the fourth year is $8,192, and the depreciation for the fourth year is $2,048.

Sum of the Years Digits Depreciation

This method of depreciation also accelerates the depreciation schedule relative to straight-line depreciation. Most assets lose value faster in the first few years. It is well known that the first mile that you drive a car off the dealer's lot costs you thousands of dollars in depreciated value. The other side of the coin is that there is not much difference in the price of a 1998 Honda Civic and a 1997 Honda Civic.

Practical Math Problem

7.6 Sum of the Years Digits Depreciation

A drill press is purchased for $6,000. It has an expected life of 8 years. What is the book value at the end of the 5th year, and what is the depreciation for the 5th year, as computed by the sum of the years digits method?

➤ C = cost of equipment

➤ N = number of years of expected life

➤ n = number of years of depreciation

➤ D = depreciation for year n

➤ B = book value after n years

$$B = C(1 - \frac{n(2N + 1 - n)}{N(N + 1)})$$

$$D = C(\frac{2(N + 1 - n)}{N(N + 1)})$$

The book value at the end of the fifth year is $1,000, and the depreciation for the fifth year is $666.67.

Social Security

We now come to the example of a practical math question that we first mentioned in Chapter 1: at what age should you take Social Security?

This is not a question with a simple answer because one of the key factors in your decision is how long you think you will live. Of course no one can be certain of how long they will live, but hopefully you will live a long time.

Your tax situation is also an issue. The government takes a percentage of your Social Security payments. If you are currently working but can claim Social Security, you might want to consider deferring it for two reasons:

> ➤ The payments to which you are entitled will be higher

> ➤ If you wait until after you retire, you will probably be in a lower tax bracket, which will also increase the net amount that you are paid

Another factor is the viability of the Social Security system. The Social Security system is expected to be viable at least until 2042, so this does not at the moment appear to be a problem. However, as everyone knows, all levels of the government are financially strapped. At the very least, you can probably expect that the age at which you can first claim Social Security (currently sixty-two) will be increased. The system may be restructured or even scrapped; no one can say.

So if you're reading this book and you're thirty years old, you might as well move on to the next topic. However, if you're considering filing for Social Security benefits soon, it would be a good idea to continue reading.

There is one aspect to Social Security that can be calculated with a substantial amount of accuracy. This is the total amount of physical dollars you will receive. As we have seen in the case of house payments, what counts is the present value of the house payments you make. In this case, what counts is the Social Security payments you receive, not the actual total dollar amount. Adding to the difficulty is that some years the payments from Social Security increase due to cost of living adjustments (COLAs).

Nonetheless, even with all those uncertainties, the following is a calculation that everyone who is considering receiving Social Security should make. Fortunately, it's a fairly easy calculation to make.

Practical Math Problem

7.7 Social Security

If you retire at age 62, you will receive a monthly Social Security check of $1,200. If you retire at age 66½, you will receive a monthly Social Security check of $1,500. What age will you be when the total income from retirement at 66½ exceeds the total income from retirement at 62?

➤ r = your age in years at early retirement (62 in this problem)

➤ m = monthly income from early retirement

➤ R = your age in years at later retirement

➤ M = monthly income from later retirement (66½ in this problem)

➤ A = your age when total income from late retirement exceeds total income from early retirement

$$A = R + \frac{m(R - r)}{M - m}$$

You will be 84½ years old when the total income from electing either plan is the same.

It's important to realize that this number is an estimate, due to the large number of factors above. If you are in good health, your family has a long history of longevity, you are currently working and either don't need or have a current use for the extra income, it would probably be the wise choice to defer taking Social Security. However, once you make this calculation, you are in a better position to make an intelligent decision—and that's one of the main purposes of practical math.

Keeping Up with Inflation

You may not always be able to tell if you are keeping up with the Joneses, but you can tell if you are keeping up with inflation.

The U.S. Government is the repository of a large amount of statistical information. Much, possibly most, of this information is publicly available, and some of it can be very helpful. The Consumer Price Index (CPI) is one of these pieces of statistical information. It is a good guide to how rapidly prices are increasing. Websites come and go, and so the policy adopted in this book will be to suggest phrases to type into a search engine to locate helpful data. In this case, it will come as no surprise that you should type *Consumer Price Index* into a search engine.

Practical Math Problem

7.8 Keeping Up with Inflation

Your after-tax income increased from $30,000 in 2009 to $30,800 in 2010. Did this keep up with inflation?

➤ A = after-tax income at end of period (2010 in example)

➤ a = after-tax income at beginning of period (2009 in example)

➤ C = Consumer Price Index at end of period (2010 in example)

➤ c = Consumer Price Index at beginning of period (2009 in example).

➤ I = ratio of after-tax increase to CPI increase

$$I = \frac{A}{a} \Big/ \frac{C}{c}$$

I > 1 outgained inflation

I = 1 stayed even with inflation

I < 1 lost ground to inflation

I > 1 outgained inflation

I = 1 stayed even with inflation

I < 1 lost ground to inflation

At the time of this writing, the CPI for October 2009 was 216.330 and the CPI for October 2010 was 218.803. This was the latest data available. The number I computed above would be 1.015, indicating a slight gain against inflation. This means that your income buys slightly more in the way of goods and services this year than it did last year.

The CPI has a base of 100 for the years 1982-1984, and a CPI in 2010 of 218 indicates that prices have risen approximately 110% since that time.

This problem can be used as a minor component in the decision as to whether you should take Social Security early or late. If you are losing ground to inflation, there is more pressure to take Social Security early.

PART THREE

Math for Living

CHAPTER 8

Everyday Problems

In This Chapter

➤ Basic ratio and proportion

➤ Time needed for doing tasks

➤ Saving money on electricity and gasoline

Can I Skip This Chapter?

This chapter can be used by practically everyone because it deals with everyday problems that occur in and around the home and because those problems involve time and money.

It is said that the majority of accidents that occur happen in the home. That's not because the home is such a dangerous place, but simply because that's where you spend most of your time.

Similarly, many of the problems for which you need practical math—anywhere from really simple, annoying stuff to some fairly difficult questions—occur in and around the home. A large portion of this book is devoted to these problems. In later chapters we'll look at some of the problems that occur in specific areas of the home, but this chapter deals mostly with general day-in, day-out problems.

We'll start off with what is certainly one of the most common problems: changing the size of a recipe.

➤ a = amount of ingredient in recipe

Practical Math Problem

8.1 Changing the Size of a Recipe

A recipe for meat loaf calls for 3 pounds of hamburger and serves 8. How much hamburger will be needed to serve 12?

> ➤ n = number served in recipe
>
> ➤ N = number desired to be served
>
> ➤ A = amount needed

$$A = \frac{N}{n}a$$

4½ pounds of hamburger are needed to make enough meat loaf to feed twelve.

Hamburger can be bought—and used—in almost any conceivable amount. Digital scales are now capable of weighing things to hundredths of a pound, and when we go to the store and ask for 3 pounds of freshly ground hamburger, the butcher can probably get the scale to register 3.00 pounds. If we're buying 3 pounds of shrimp, that's probably not so easy to do, but generally we're willing to settle for 2.96 pounds or 3.03 pounds of shrimp.

However, if the above recipe is actually a recipe that makes a cake for eight using three eggs and we want to extend the recipe to serve twelve, the calculation we just made shows that we need four and a half eggs. There are four possible solutions to the problem in this case; you pick the one that is best for you:

> ➤ If indeed you have a way to break an egg, divide it roughly in two, and have a use for the other half egg (such as frying it and eating it)
>
> ➤ Use only four eggs and make the cake with a little less than the ideal amount of egg. Probably no one will notice.
>
> ➤ Use five eggs and make the cake with a little more than the ideal amount of egg. Probably no one will notice.
>
> ➤ Extend the size of the recipe to serve sixteen, which requires doubling the amount of each ingredient in the recipe. You can generally find someone to take the extra four helpings of cake off your hands.

The last solution brings us to an extremely important idea: the concept of ratio and proportion. This is a practical math book, but there are so many problems around the house that are solved through a simple knowledge of ratio and proportion that it's worth spending a small amount of time on the basic idea. You'll also see a lot of examples of problems that can be solved using this one idea.

Ratio and Proportion

Ratios and proportions are one of the most important tools in the mathematician's problem-solving kit, and it's one you already understand from the scenario just discussed. If you have a recipe that serves eight and you want to use the same recipe to serve sixteen, you simply double the amount of each ingredient in the recipe. Here's an example of the same idea in reading a map.

Admittedly, this problem is somewhat of a slam dunk, as long as you understand the idea

Practical Math Problem

8.2 Reading a Map

The scale of a map is 1 inch to 50 miles. What does a distance of 5 inches on the map represent?

from problem 8.1, but we'll go through the formalities anyway:

➤ m = map distance of example (1 inch in this problem)

➤ d = actual distance of example (50 miles in this problem)

➤ M = given map distance (5 inches in this problem)

➤ D = real world distance

$$D = \frac{M}{m}d$$

The actual distance corresponding to 5 inches on the map is 250 miles.

Hopefully, you've noticed that the formula that solves problem 8.2 is essentially the same formula that solves problem 8.1. The only difference is that problem 8.2 uses D and d in place of A and a, and M and m in place of N and n.

Many of the formulas in this book are the result of setting up a proportion, which is the equality of two ratios, and solving the resulting equation. A ratio is the comparison of two numbers by a division or an implied division (a division you have not yet carried out). On the map in problem 8.2, it would be expressed as 1 inch:50 miles.

Whether or not ratios include units is extremely important. It is also possible to express ratios without using units in case the units are the same. The map ratio in problem 8.2 could have been expressed as 1:3,168,000, because there are 3,168,000 inches in 50 miles, but it isn't as easy to read or understand as 1 inch:50 miles.

In problem 8.2, the form of the proportion in English is:

m is to M as d is to D.

Substituting the two ratios, we now express this mathematically.

m:M = d:D

Using the fact that a ratio is a division, the above proportion translates to the equation which follows.

$$\frac{m}{M} = \frac{d}{D}$$

The solution of this equation is given in the formula preceding the answer. This formula can be regarded in the more general form:

$$\textbf{New Value of Quantity 2} = \frac{\textbf{New Value of Quantity 1}}{\textbf{Old Value of Quantity 1}} \times \textbf{Old Value of Quantity 2}$$

It would be impossible to give a complete list of all the problems that this generic formula solves. More are being found every day, but here are some examples of common problems that are found in and around the home.

➤ If 1 pound of sugar costs 80¢, how much does 5 ounces of sugar cost?

➤ If six screws are required to build a birdhouse, how many will be required to build three birdhouses?

➤ If it takes 2 hours to make a trip by plane that requires 12 hours by train, how long will it take to make a trip by train that requires 2 hours and 45 minutes by plane?

The first two of these problems have solutions that are guaranteed; 5 ounces of sugar cost 25¢ and eighteen screws will be required to build three birdhouses. The last problem, however, comes with the standard mathematical disclaimer of "all other things being equal." For instance, train travel is generally not affected by wind conditions, but plane travel certainly is. If the 2-hour plane trip is from east to west but the 2-hour-and-45-minute plane trip is from west to east, the answer one gets from solving the ratio (16½ hours) is only an approximate answer and one that may miss the mark considerably. However, at least it gives you some idea of how long the train trip will take.

Time Needed for Household Tasks

Let's suppose you know how long it takes to do a job. One of the most useful things to do with this information is to project how long it will take to do similar jobs. This is one of the major uses of ratio and proportion. We know that if it takes 1 hour to make a 40-mile trip on the freeway, if we have to go 60 miles it will take somewhere in the vicinity of 1½ hours.

However, there are time projections that are not directly handled by the simple method of ratio and proportion that we have already discussed.

Two People, One Job

There are a number of problems that fit into the following framework. The time required by each of two people to complete the job if they worked on their own is known. The question is how long it would take them if they worked together to complete the job.

Practical Math Problem

8.3 Time Needed for a Two-Person Job

John can wash the dishes in 30 minutes if he does them by himself. Sue can wash the dishes in 20 minutes if she does them by herself. How long will it take both of them working together to wash the dishes?

➤ t = time needed by one person

➤ s = time needed by the other person

➤ T = time required for both

$$\mathbf{T} = \frac{\mathbf{st}}{\mathbf{s+t}}$$

It takes 12 minutes for them to wash the dishes. Of course, *working together* here means that they don't spend time arguing over who washes the frying pan, and that when they work together each works at the same rate they would if they worked separately. These may not be valid assumptions, but they are necessary to come up with an answer for the problem, and the answer of 12 minutes will certainly be a good estimate.

There are some related problems in which the answer is almost certain to be extremely close to what actually happens. If one has two pipes leading into a swimming pool, and one pipe can fill the pool in 30 minutes and the other can fill the pool in 20 minutes, if both pipes are turned on simultaneously it will take 12 minutes to fill the pool.

Several People Working on Repetitive Tasks

Christmas gives many families an opportunity for adults—and children—to work together to complete a job through the repetition of a simple task. One such job is addressing Christmas cards. The number of people working on the job can vary, and the number of Christmas cards that need to be addressed can change as well.

Practical Math Problem

8.4 Time Needed to Address Christmas Cards

If three people can address 20 Christmas cards in 50 minutes, how long will it take 4 people to address 35 Christmas cards?

➤ p = number of people in given information

➤ c = number of Christmas cards in given information

➤ t = time needed in given information

➤ P = number of people available

➤ C = number of Christmas cards that must be addressed

T = time needed to address C Christmas cards using P people

$$T = \frac{tpC}{Pc}$$

It takes between 65 and 66 minutes for the four people to address thirty-five Christmas cards.

This problem actually consists of two successive applications of the ideas of ratio and proportion. If three people can address twenty Christmas cards in 50 minutes, it would take four people three-quarters of 50 minutes, or 37.5 minutes, to address twenty Christmas cards. Since the ratio of 35 to 20 is the same as the ratio of 1.75 to 1, it will take 1.75 x 37.5 minutes for four people to address thirty-five cards.

Another Trip to the Kitchen

We're going back to the kitchen to look at some other common problems. Mixing is something that often takes place in the kitchen—and elsewhere.

Mixing Two Mixtures with Different Percentages

This is an example of a problem that can occur not only in the kitchen, but when medicines are mixed as well.

Practical Math Problem

8.5 Mixing Two Mixtures

A 2-quart pitcher of fruit punch is 60% fruit juice. How many quarts of fruit punch that is 80% fruit juice must be mixed with this in order to have a mixture that is 75% fruit juice?

➤ v = volume of original material

➤ p = percentage of key ingredient in original material

➤ q = percentage of key ingredient in added material

➤ P = desired percentage of key ingredient in final mixture

➤ V = needed volume of added material

$$V = \frac{v(p - P)}{P - q}$$

In order to create a mixture that is 75% fruit juice, 6 quarts of 80% fruit juice must be added.

Alternative Forms

$$q = P + \frac{v(P - p)}{V}$$

This formula is used to answer questions such as "What does the percentage of alcohol have to be in a mixture for which 4 quarts are to be added to 2 quarts of a mixture that is 10% alcohol in order to create a mixture that is 20% alcohol?" The answer to this question is 25%.

These formulas have numerous uses outside the kitchen. Mixologists will note that instead of mixing fruit punch, it is possible to mix shakers of martinis that have been mixed at different gin-to-vermouth ratios.

More importantly, these formulas can be used to tackle the problem of what the percentage of active ingredient is in a solution of two mixtures with different percentages of that active ingredient. This can be useful in mixing drugs or other chemicals.

How Long Will it Take Soup to Cool?

There is a saying that a watched pot never boils. The other side of the problem is equally frustrating: waiting for soup to cool so that it doesn't burn your tongue.

Most of the problems so far presented in this book come from algebra or subjects related to algebra. The question of how long it takes things to cool was actually first investigated by Isaac Newton using techniques from calculus, which he invented.

Practical Math Problem

8.6 When Will the Soup Cool?

After it has been placed in the microwave for heating, the temperature of a bowl of soup is 160°. It takes 5 minutes for it to cool to 140°. How long will it take to cool to 120° if the temperature of the room is 70°?

➤ R = room temperature

➤ T_1 = temperature of soup at start (1st measurement)

➤ T_2 = temperature of soup at 2nd measurement

➤ s = time between 1st and 2nd measurements

➤ T_3 = temperature of soup at 3rd measurement

➤ t = time between 1st and 3rd measurements

$$t = s \ \ln\left(\frac{T_3 - R}{T_1 - R}\right) \Big/ \ln\left(\frac{T_2 - R}{T_1 - R}\right)$$

It will take about 11.7 minutes for the soup to cool from the time it was removed from the microwave, or about 6.7 more minutes from the time it cooled down to 140°. Even though the measurements given were in degrees Fahrenheit, the formula is equally applicable if all temperature measurements are made in degrees Celsius.

This formula is widely used in forensic medicine and in determining the time of death, as in the following example:

At 1 am, the body is discovered in the conservatory with severe blunt-force trauma to his head. His body temperature is 90°. By 1:45 am, the body temperature has declined to 86°. The temperature of the conservatory is kept at a constant 78°. What time was he killed (bonus points for who did it and with what)?

The previous formula can be used with T_1 denoting the temperature at the time of death—in most cases that's 98.6°. Here it is t that is known (45 minutes), and the above formula solved for s. We'll save you the trouble of doing this problem The death took place at exactly midnight, and devotees of Clue know that the murder is often committed in the conservatory, by Colonel Mustard, with the wrench. Or maybe it was alliteratively done by Professor Plum with the pipe.

The Cost of Electricity

We run most, if not all, of household appliances on electricity, and then we generally simply pay the electricity bill and forget about it. There's a lot of money spent annually on electricity, and a lot to be saved as well.

Practical Math Problem

8.7 The Cost of Electricity

Electricity costs 12¢ per kilowatt-hour, and a typical computer and monitor use about 0.3 kilowatts per hour. How much does it cost to leave the computer and monitor on for 8 hours nightly?

➤ K = number of kilowatts used by appliance

➤ H = number of hours in service

➤ C = cost per kilowatt-hour of electricity

➤ T = total cost

$$T = CHK$$

It costs about 29¢ to leave the computer and monitor on overnight. It's aggravating to have to turn the computer on in the morning and wait for it to load up, but fortunately there's an effective compromise. If you place them both in sleep mode, they only draw about 0.03 kilowatts, which cuts your costs by about 90%. This amounts to a savings of 26¢ per night— almost $95 per year! Of course, if you live in a region with cheaper electricity than Southern California you won't save as much, but you'll probably still save in the vicinity of $50 a year. Do this for a decade and you can purchase a very nice computer and monitor—and maybe even a printer.

To help you calculate your cost of electricity, here is a table of common household appliances and the approximate amount of electricity they use. These are ones that not only draw a lot of power, but also tend to be left on for long periods of time. You'll have to consult the manufacturer of the specific brand you use for the exact amount of electricity that it draws.

Appliance	kW/hr
Air Conditioner (5,000 BTU)	0.9
Air Conditioner (12,000 BTU)	1.5
Clothes Dryer	5
Electric Lights (variable)	2.5
Electric Range	12
Water Heater	4

If you spend a little while actually tracking the use of these major appliances during a few months, you'll probably find that you are leaving hundreds of dollars on the table annually due to simple failure to turn machines off. You don't need to cool a house when no one is home, and there's no need to leave the lights on while you're sleeping—at least, not all of them.

The Cost of Gasoline

Gasoline cost is a much more obvious cost than electricity. We generally don't leave the car motor running when we're not in the car, although we do leave electrical appliances on when we're not around.

Savings from Driving a Hybrid

There are two big reasons for buying a hybrid: the cost savings from not having to buy as much gasoline, and the feeling that you are doing something ecologically responsible. Since there's really no way to put a dollar figure on the latter, let's take a look at your annual gasoline savings from driving a hybrid.

Practical Math Problem

8.8 Gasoline Savings from Driving a Hybrid

A car with a standard internal combustion engine gets 30 miles per gallon, whereas a hybrid gets 50 miles per gallon. Gasoline costs $3.00 a gallon, and you drive an average of 12,000 miles annually. What is your annual savings on gasoline from buying a hybrid?

➤ m = miles per gallon with standard engine

➤ M = miles per gallon with hybrid

➤ G = number of gallons purchased annually

➤ C = cost per gallon of gasoline

➤ S = annual gasoline savings from buying a hybrid

$$S = CG(\frac{1}{m} - \frac{1}{M})$$

The annual savings on gasoline from driving a hybrid is $480.

From a financial standpoint, this is a prime factor in considering the cost savings of buying a hybrid, but there are other factors in play as well. Nonetheless, since hybrids generally cost $5,000 or more than comparable traditional cars,, and given the current gas prices, the cost savings from buying a hybrid don't seem competitive—most people do not keep a car for ten years. Higher gas prices or standard engines getting lower gas mileage would undoubtedly impact this decision.

Driving to Save Money on a Purchase

Many of us drive to a supermarket that's a little further away because the price of chicken is 30¢ per pound lower at the further supermarket. But do we really save money by doing so?

Practical Math Problem

8.9 Driving to Buy at a Cheaper Price

Gasoline costs $3.00 a gallon and your car gets 25 miles a gallon. How far would you go to save $30 on the purchase of a digital camera?

> ➤ S = amount of savings

> ➤ g = cost of a gallon of gasoline

> ➤ m = miles per gallon

> ➤ D = maximum one-way distance to alternate purchase site

$$D = \frac{Sm}{2g}$$

You would break even by driving 125 miles to purchase the camera at a discount of $30. This assumes that you do not include the cost in time of driving since you are driving during your spare time.

CHAPTER 9

Automobile Usage and Performance

In This Chapter

➤ Fuel economics

➤ Under the hood

➤ On the road

Can I Skip This Chapter?

There is no question that the automobile ranks right up there on the list of inventions that have changed civilization. It's certainly changed America; we are a car-loving, car-dependent culture. A car is a major investment as well as a symbol of adulthood. There are some people whose lives center around their cars, and we're not simply referring to NASCAR drivers.

This chapter deals with the economics of automobile usage, so if you own a car, you'll want to read this chapter. It also discusses the mathematics of automobile performance, which is useful mainly for those who spend time maintaining automobiles.

There's a lot of practical math associated with cars. Some of it will be explored in this chapter, but we'll see other examples in later chapters when we look at some of the engineering and physics aspects of engines and motors.

Fuel Economics

We're going to bring back problem 8.8 in a slightly different guise because not only is there an important point to be made that hasn't been made yet, it serves as a good lead-in to the next topic.

Practical Math Problem

9.1 Savings from Better Mileage

What is the annual savings from driving 12,000 miles per year with a car that gets 12 miles per gallon as opposed to a car that gets 10 miles per gallon, if fuel costs $3 per gallon?

- ➤ D = annual distance traveled
- ➤ C = fuel cost per gallon
- ➤ M = more miles per gallon
- ➤ m = fewer miles per gallon
- ➤ S = annual savings

$$S = CD\left(\frac{1}{m} - \frac{1}{M}\right)$$

The savings in this instance is $600. To refresh your memory, when comparing the standard engine, which got 30 miles per gallon, to the hybrid, which got 50 miles per gallon, with the same annual mileage and gasoline cost, the annual gas savings form the hybrid was $480.

There's a trap here, so we'll point it out to help you avoid it. The gas mileage from the hybrid was 66 2/3% better than the gas mileage from the standard engine in that comparison. In problem 9.1, a gas mileage of 12 miles per gallon is only 20% better than a gas mileage of 10 miles per gallon, but the annual savings is greater with the 20% increase in mileage in this situation than with the 66 2/3% savings with the hybrid.

The trap is that it isn't so much the percentage increase in miles per gallon that matters; it's the actual miles per gallon that counts. This is often unmentioned by the manufacturers of fuel additives, who will tout the percentage increase in miles per gallon that their additive produces. So let's take a closer look at the mathematics of the situation.

Fuel Additives

The first half of the twentieth century featured rumors of a mysterious carburetor that was being suppressed by the oil industry because it enabled cars to get 100 miles per gallon.

Ever since then, and perhaps even before, people have been seeking fuel additives that when added to a gallon of gasoline increased gas mileage (and also made the engine run more smoothly, at least according to the makers of the additive). Let's look at the economics of fuel additives.

Practical Math Problem

9.2 Savings from Fuel Additives

You own a car that currently gets 24 miles per gallon. Gas costs $3.00 per gallon. A fuel additive costs $2.00 per ounce, and one ounce must be added to 8 gallons of gasoline to be effective. What mileage must the car get with the fuel additive in order to justify the cost of the additive?

➤ c = cost of 1 ounce of fuel additive

➤ C = cost of 1 gallon of gas

➤ m = current mileage in miles per gallon

➤ n = number of gallons treated with 1 ounce of fuel additive

➤ M = improved mileage with fuel additive in miles per gallon

$$M = m\left(1 + \frac{c}{Cn}\right)$$

You need to get 26 miles per gallon in order to justify the cost of the fuel additive. Of course, you want to make sure that using the additive does not harm your engine; otherwise what you gain in gas savings may be lost in repair costs.

Under the Hood: Engine Parameters and Vehicle Modification

Internal combustion engines are run by the reciprocating motion of a piston in a cylinder that is transformed to rotary motion, which drives the wheels of the car. The engine configuration is often specified by how the cylinders are aligned. The V8 engine, which is still popular, consists of eight cylinders mounted four on each side of a *V*.

Bore, Stroke, and Displacement

The key engine parameters are the bore, stroke, and displacement of the engine. The bore is the diameter of an individual cylinder. The stroke is the distance the piston travels. The displacement is the total volume swept out by all the pistons from the top to the bottom of the piston movement.

Practical Math Problem

9.3 Bore, Stroke, and Engine Displacement

What is the displacement of a six-cylinder engine with a 4-inch bore and a 3.5-inch stroke?

➤   B = bore

➤   S = stroke

➤   N = number of cylinders

➤   D = displacement

$$D = \frac{\pi}{4} NB^2 S$$

The engine displacement is 263.9 cubic inches.

If you are familiar with the formula for the volume of a cylinder (this will be covered in Chapter 24), you will realize that this is simply the number of cylinders multiplied by the volume of each cylinder.

A little algebra enables us to retrieve each of the three variables in terms of the other two.

Alternative Forms

$$B = \sqrt{\dfrac{4D}{\pi NS}}$$

This formula answers questions such as "What is the bore of a four-cylinder engine with a displacement of 250 cubic inches and a stroke of 3.75 inches?" The answer to this question is approximately 4.6 inches.

$$S = \dfrac{4D}{\pi NB^2}$$

This formula answers questions such as "What is the stroke of an eight-cylinder engine with a displacement of 408 cubic inches and a bore of 4 inches? The answer to this question is approximately 4.06 inches.

Compression Ratio

An engine's compression ratio is the ratio of the largest volume to the smallest volume of the combustion chamber. It is one of the fundamental parameters that describe the engine. High compression ratios mean that the fuel-air mixture is highly compressed. This results in faster and more complete burning. High compression ratios formerly came with a liability; lower octane fuel would result in engine knocking. But this has been effectively eliminated in the newer cars.

Practical Math Problem

9.4 Compression Ratio

An engine has a 4-inch bore, a 3.5-inch stroke, a measured chamber volume of 4.27 cubic inches, and a gasket thickness of 0.05 inches. What is its compression ratio?

➤ B = bore

➤ S = stroke

➤ G = gasket thickness

➤ V = chamber volume (this must be measured manually, as the chamber is irregular in shape)

➤ C = compression ratio

$$C = \frac{\pi B^2 S}{4V + \pi B^2\, G} + 1$$

The compression ratio is 9.98 to 1.

A standard way to improve a car's compression ratio is by milling the cylinder heads. A small amount of milling can actually increase the compression ratio significantly.

Practical Math Problem

9.5 Milling the Heads

By how much must the cylinder heads be milled on an engine with a 3.5-inch stroke in order to increase the compression ratio from 9.6 to 10.5?

> ➤ C = old compression ratio

> ➤ N = new compression ratio

> ➤ S = stroke

> ➤ M = amount to be milled

$$M = \frac{N - C}{(N - 1)(C - 1)}S$$

The heads should be milled by 0.039 inches.

Engine Horsepower

The standard measure of the power of an engine—the rate at which it can use energy—is the number of horses that it has under its hood.

Practical Math Problem

9.6 Horsepower of an Auto Engine

What horsepower is delivered by a six-cylinder engine in which a single cylinder has a radius of 3 inches, a stroke of 8 inches, a crank that makes 250 revolutions per minute, and a mean effective pressure of 100 pounds per square inch?

➤ C = number of cylinders

➤ P = mean effective pressure in pounds per square inch

➤ L = length of stroke in inches

➤ R = radius in inches

➤ N = number of revolutions per minute

➤ H = engine horsepower

$$H = \frac{\pi CPLNR^2}{198,000}$$

The engine delivers approximately 171 horsepower.

Taking Your Car on the Road

Now that you've figured out whether or not a fuel additive is a good deal and have modified your car to improve its compression ratio (or refrained from doing so), it's time to take it out on the road.

Bad things can happen to your car on the road. It could get into an accident—but we've covered the practical math aspect of that in problem 4.2, where we discussed which deductible to take when buying insurance. But there are other pitfalls to watch out for.

Avoiding Speeding Tickets

Another bad thing that could happen besides getting into an accident is that you could get a ticket for speeding. Yes, a large portion of the responsibility for speeding is in your hands (or foot)—after all, it's your foot that's on the accelerator. However, there is a possibility that your speedometer could lead you astray by registering a legal speed when in fact you are driving too fast. Some highways have measured miles so you can check the accuracy of your speedometer by doing your best to maintain a constant speed, preferably 60 miles per hour. Sometimes traffic conditions will not allow that, so you maintain whatever constant speed you can. Check your watch at the start and at the end of the measured mile.

Practical Math Problem

9.7 Checking Your Speedometer

It takes you 87 seconds to complete a measured mile during which your speedometer registered a constant 40 miles per hour. What is your true speed, and what percentage of your measured speed is your true speed?

> ➤ C = constant speed over measured mile

> ➤ T = number of seconds to complete measured mile

> ➤ S = true speed of car

> ➤ P = true speed percentage of measured speed

$$S = \frac{3600}{T}$$

$$P = 100\frac{S}{C} = \frac{360,000}{TC}$$

Your true speed is 41.38 miles per hour, and your true speed is approximately 103.4% of your measured speed. That 103.4% is a good approximation of your true speed as a percentage of your measured speed, no matter how fast you are going.

Could this get you a speeding ticket? Quite possibly. In Los Angeles, the nominal speed limit on many freeways is 70 miles per hour. In good economic times, the police were known to allow you a 5-mile-per-hour halo and would not give you a ticket unless you were doing 75. It's not clear that such is the case in bad economic times, as they may be giving more tickets to "improve public safety" as well as police department revenue. If your speedometer says 73, that extra 3.4% will boost your speed to slightly more than 75 miles per hour.

There's another way that your speedometer may not be telling it like it is. If you are one of those individuals who is enamored of putting tires with larger diameters on your cars, your speedometer will again lie to you. The speedometer computes your speed as a function of the number of revolutions the axle makes, but that is calibrated to standard tires on your car. The distance your car travels in a single revolution of the axle is equal to the circumference of the tire. Increase the tire diameter, and the circumference increases, so you travel a greater distance during a single revolution.

Practical Math Problem

9.8 Tire Diameter Effect on the Speedometer

If the tire diameter is changed from 36 inches to 38 inches, what is the true speed when the speedometer reads 60 miles per hour?

➤ t = old tire diameter

➤ T = new tire diameter

➤ S = speedometer reading

➤ V = true speed

$$V = \frac{T}{t}S$$

The true speed is 63.3 miles per hour; an error of more than 5%.

Be Sure to Set Your Brakes on a Hill!

A staple of film and television comedy is watching a parked car on a hill with brakes that have not been set or wheels that have not been insufficiently curbed rolling downhill. It's funny when it's on the screen, but not nearly so funny in real life.

Although angles are generally measured in degrees or radians, the steepness of a hill is generally given as a percentage when it is posted on signs.

Practical Math Problem

9.9 Force Needed to Hold Car on Hill

A 4,000-pound car is parked on a hill at a 9% angle. How much force is required to prevent the car from rolling down the hill?

➤ W = weight of car

➤ p = angle of hill (expressed as grade of p percent)

➤ F = force required

$$F = \frac{Wp}{\sqrt{p^2 + 1}}$$

It requires 358.6 pounds of force to keep the car from rolling down the hill. Even if you're a power lifter, set the brakes and curb the wheels!

When You Need a Tow

Every so often, either you or a friend needs a tow, and a towing service is either not available or too expensive, so you decide to do it yourself. Be careful; accelerating too rapidly can break the tow rope (although probably not a tow chain).

You may be a little uncomfortable with the next problem because the key parameters are measured using the unfamiliar metric system. There's a later chapter on conversions to handle this, but there's a good reason for doing problems using the metric system, which will be discussed after the problem.

Practical Math Problem

9.10 Maximum Acceleration in Towing a Car

A rope will break if stretched with a force of 3,000 newtons or more. What is the maximum acceleration with which it can tow a 1,500-kilogram car on level ground?

> ➤ M = mass of car in kilograms

> ➤ F = breaking strength of rope in newtons

> ➤ A = maximum acceleration in meters per second

$$A = \frac{F}{M}$$

The maximum acceleration is 2 meters per second per second, which is about 6.56 feet per second per second, or 4.5 miles per hour per second. If you happen to be into drag racing, an acceleration of 4.5 miles per hour per second will result in a 0 to 60 acceleration time of about 13 1/3 seconds, which certainly won't win you any prizes—except an unbroken tow rope.

If force and mass are both measured in pounds, the formula A = 32 F/M will give the acceleration in feet per second, but it is important to ascertain that the rope-breaking strength is in pounds of force rather than pounds of mass. It is probably safer to do this problem in the metric system, and then convert the meters per second per second into miles per hour per second. Seeing the number in miles per hour per second will give you a good feel for how fast you can accelerate your car.

Finally, it's a good idea to know how far you will travel if you have to suddenly apply the brakes.

Practical Math Problem

9.11 Distance Traveled After Braking

A car traveling 50 miles per hour brakes suddenly and comes to a complete stop in 6 seconds. How far does it travel between the time the brakes are applied and the time it comes to a complete stop?

➤ v = initial velocity in miles per hour

➤ T = time in seconds to come to a complete stop

➤ D = distance in miles before coming to a complete stop

$$D = \frac{vT}{7200}$$

The car will travel 1/24 of a mile, or 220 feet. It's not a bad idea to get some idea of these numbers before you have to use them in an emergency situation. The rule of thumb that you should allow one car length between you and the car in front of you for every 10 miles per hour you are going is still a pretty good one to follow.

CHAPTER 10

The World Around Us

In This Chapter

➤ Taking a trip

➤ Travel expenses

➤ Distances, heights, and times

Can I Skip This Chapter?

Don't skip this chapter if you're interested in the cost of travel or measuring distances, heights, and times.

By and large, most people love to travel, to get away from the humdrum and familiar and see different places. We are remarkably fortunate to live in an age where we can hop in our car and take a relaxed and pleasant vacation within easy driving distance of home, or board a boat or plane and travel practically anywhere on earth. The circumference of the earth is about 25,000 miles, so the furthest that any two points on earth can be from one another is about 12,500 miles. It would take 21 hours in a jet plane capable of doing 600 miles per hour to span that distance.

Yes, it's a little frustrating that it takes more time to get to a New York airport from downtown New York and go through security than it does for the plane to go from New York to Chicago. But the frustration is nothing when we consider that little more than two centuries ago, such journeys were hazardous ventures, and little more than a century ago, a train capable of 60 miles per hour or so was the fastest form of transportation.

When we go places, we are interested in knowing the times, distances, and costs. When we go to Paris, for example, we want to know how long it will take, how far away it is, how long it will take us to get there, and how tall the Eiffel Tower really is. It's not just a trip to Paris; even a local trip often raises some of the same questions.

This is the first chapter that has two features not yet seen before: trigonometry functions and diagrams. Of course, it isn't necessary for you to review trig; just stick the numbers into the calculator or spreadsheet to get the answer. The diagrams in this chapter consist of straight lines with labels to indicate variables. This is rather primitive, but many problems are illuminated by seeing a picture, no matter how rudimentary.

A Matter of Time

We live in a fast-paced world, and almost all of us are preoccupied by saving time. Connecting with 4G is faster than with 3G, the computer never downloads fast enough, and we always seem to be in a hurry. One of the places where we often try to make up for lost time is on the road—and maybe that's not such a good idea.

Practical Math Problem

10.1 Time Saved by Going Faster

How much time can you save on a 200-mile trip by averaging 60 miles per hour rather than 50 miles per hour?

➤ D = distance traveled

➤ M = faster speed

➤ m = slower speed

➤ T = time saved

$$T = \frac{D}{m} - \frac{D}{M}$$

You can save 2/3 of an hour, or 40 minutes. That's a significant amount of time, but 200 miles is a long road trip, taking the better part of a day.

Try crunching the numbers for a 20-mile trip and look at the difference between going a fairly safe 60 and pushing the speed envelope at 75. You save 4 minutes! Is it really worth the risk?

Making Up for Lost Time—and Money

We've all found ourselves in this predicament: we started late, or ran into unexpected traffic, and now we're behind schedule. What do we need to average for the remainder of the trip in order to make up for the time we've lost?

Practical Math Problem

10.2 Making Up for Lost Time

You wish to average 50 miles per hour on a 300-mile trip, but unfortunately you only averaged 40 miles per hour for the first 2 hours. How many miles per hour must you average on the remainder of the trip to make up for lost time?

➤ D = trip distance

➤ A = desired average speed

➤ H = time taken for first portion of the trip

➤ s = average speed maintained on first portion of the trip

➤ V = average speed needed on remainder of trip

$$V = A\frac{D - sH}{D - AH}$$

You'll need to average 55 miles per hour for the remainder of the trip.

Sometimes You Don't Need a Calculator

This may look like a messy calculation, but there are two easy ways to do it without a calculator. Since you averaged 40 miles per hour for the first 2 hours, you were 10 miles per hour behind your desired average over the first 2 hours, or a total of 20 miles per hour. Your plan to average 50 miles per hour for a 300-mile trip means that the trip should take 6 hours, and you've already used up two of those hours. That leaves 4 hours to make up those 20 miles, and you can do that by averaging five miles per hour faster than the desired average of 50 miles per hour for the entire trip. Do the math and you get the same answer as you did with the formula: $50 + 5 = 55$.

There's another, even simpler, way to do it. Averaging 40 miles per hour for the first 2 hours, you drove 80 miles. That leaves 220 miles. You have 4 hours to drive those 220 miles, so you have to average 55 miles per hour.

The formula used in problem 10.2 actually does double duty. We can use it to get back on track after we've overspent our budget.

Assume that we've budgeted $3,000 for a two-week vacation, which is on the order of $200 a day. We spent $300 a day for the first four days. How do we get back on track with the budget?

It's basically the identical calculation as in problem 10.2 except that dollars take the place of total miles and days take the place of hours. Our total allotted budget is $3,000, and in four days of averaging $300 per day, we've spent $1,200. That leaves $1,800, so since we have ten days remaining, we need to average $180 per day to stay within budget.

Point of No Return

Ever since travel began, voyagers have often asked the question, "Should we turn back?" Usually, this question was asked because supplies were running low. Nowadays that's generally not a problem, but it still can be useful to know that there's a point on your journey where there's no turning back. This moment in time—or location in space—is known as the point of no return.

Practical Math Problem

10.3 Point of No Return

The distance from Los Angeles to Tokyo is 5,500 miles. A jet can fly 600 miles per hour in still air, but there is a west-to-east headwind of 50 miles per hour. How long after takeoff from Los Angeles will it be until the plane reaches the point of no return, where it takes the same amount of time to continue to Tokyo as it does to return to Los Angeles?

➤ D = distance between the two cities

➤ v = velocity of plane in still air

➤ w – wind speed

➤ t = time to point of no return

$$t = \frac{D(v + w)}{2(v(v - w))}$$

It will take 5.42 hours—roughly 5 hours and 25 minutes—to reach the point of no return.

If the plane takes off with a tailwind, use the negative of that number for w. For instance, if there is a tailwind of 50 miles per hour, w = -50.

There is another version of the point of no return problem: how far can you safely venture with a limited supply of fuel?

Practical Math Problem

10.4 How Far Will Your Fuel Take You?

A boat with a 12-gallon tank capacity gets 30 miles per gallon inbound (with the tide) and 20 miles per gallon outbound (against the tide). How far out can it safely venture?

> ➤ m = miles per gallon outbound

> ➤ M = miles per gallon inbound

> ➤ G = tank capacity in gallons

> ➤ D = maximum safe travel distance

$$D = G \frac{mM}{m + M}$$

The maximum distance for safe travel is 144 miles. However, that's an absolute maximum. Build in a margin for safety to guard against unforeseen setbacks. Instead of going 144 miles out, set a limit of 120 miles, slightly more than 80%.

There are at least two good reasons for doing so:

> ➤ If things go wrong, there's no guarantee you'll be rescued

> ➤ Even if you are rescued, in these tight fiscal times many communities are charging for services such as these, and you'll probably have to pay an arm and a leg

Allowing for a Current—or a Headwind

We don't always travel on still waters or in smooth air. Often there's a current that affects the situation by making our journey faster or more fuel-efficient in one direction, and more difficult in the other direction.

This was the case in problems 10.3 and 10.4: we knew what the conditions were and could make our computations using that information. However, sometimes you can make the trip safely and use that information to determine what the conditions were.

Air Speed—Wind Speed (or Upstream—Downstream)

Practical Math Problem

10.5 Determining Air Speed and Wind Speed

A plane traveling west from Denver makes the 1,000-mile trip to Los Angeles in 2 hours. Going east 1,800 miles to Orlando takes 3 hours. Assuming that the wind conditions are the same in both directions, what is the speed of the plane and what is the wind speed?

➤ D = distance of eastbound flight

➤ T = time taken for eastbound flight

➤ d = distance of westbound flight

➤ t = time taken for westbound flight

➤ v = speed of plane

➤ w = wind speed

$$v = 0.5(\frac{D}{T} + \frac{d}{t})$$

$$w = 0.5(\frac{D}{T} - \frac{d}{t})$$

The plane's speed is 550 miles per hour, the wind speed is 50 miles per hour. Of course, the assumption that the wind speed remains the same in either direction could be substantially in error. However, it is likely to be correct if instead of looking at the trips from one city to two different cities, we look at the round-trip information between one city and another.

This is also useful for determining the current in a river. Make a round trip going upstream in one case and downstream in the other. If you do this on a number of days, the average of the water speed in the river is a good estimate to use in planning other river journeys.

Crossing a River Against the Current

You just want to get to the point on the other side of the river that's directly opposite from you, but there's a current flowing in the river. How do you compensate for this?

Practical Math Problem

10.6 Crossing a River Against the Current

A motorboat is capable of going 15 miles per hour in still water. A current of 4 miles per hour is flowing in a river that is ¼ of a mile wide. At what angle should the boat travel in order to land at the point on the opposite bank directly across from where it starts, and how long will the trip take?

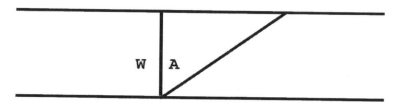

➤ W = width of river

➤ C = speed of current

➤ V = velocity of boat

➤ A = launch angle of boat

➤ T = time to cross river

$$A = \sin^{-1}\left(\frac{C}{V}\right)$$

Once A has been computed, we can compute T.

$$T = \frac{W}{V\cos A}$$

The boat should be launched at an angle of 15.47°. The trip will take about 1.04 minutes.

The Cost of Converting Currency

If you travel abroad, you are going to have to convert your currency into the other country's currency. Whenever money changes hands, some money gets lost in the cost of doing business.

Although you may see rates quoted online or in the paper as 1 dollar = 0.73 euros (or whatever the rate actually happens to be that day), this is not the rate you get when you try to convert your currency. If you go to a bank or other institution that converts currency, you will see the rate quoted in something like this fashion: dollars to euros 0.72-0.75. This means that if you are buying euros for your dollars, each dollar that you exchange will give you 0.72 euros. On the other hand, if you are exchanging euros for dollars, you will need to hand over 0.75 euros for each dollar you receive.

To add insult to injury, the institution that exchanges currency will charge an additional fee for ripping you off in this fashion.

Practical Math Problem

10.7 Cost of Currency Conversion

Suppose that the currency conversion rate is quoted as above: dollars to euros 0.72-0.75. What percentage is lost by converting dollars to euros and back again?

➤ A = number of foreign currency units a dollar will buy (0.72 in problem 10.7)

➤ B = number of foreign currency units needed to buy 1 dollar (0.75 in problem 10.7)

➤ P = percent lost to back-and-forth currency exchange

$$P = 100(1 - \frac{A}{B})$$

You lose 4% in a back-and-forth currency exchange (exchanging dollars to euros and back again) in the above example. That doesn't even include the extra cost of performing this transaction.

What can you learn from this before going on a trip? First, it's advisable to see how credit card transactions are handled; every credit card has a different arrangement. Second, if you have to exchange money, it would be nice to incur as few extra charges as possible. Ideally, you'd like to run out of foreign currency just as you get back to the United States. Many savvy travelers exchange money once, get more foreign currency than they need, and visit the duty-free shops in the airports or harbors just before they return to get rid of excess foreign currency.

This type of arrangement, in which you buy something at one price but sell at another, also occurs when you are buying and selling stocks. The price of a stock is quoted in what is known as a bid-asked; you might see a stock such as Microsoft quoted at 27.30-27.45; it costs $27.45 to buy a share of Microsoft, but if you want to sell the stock you can only get $27.30 per share. The bid-asked spread as a percentage of the stock price tends to decrease with higher share prices, so the cost of doing business tends to be a little lower for the higher-priced stocks. The bid-asked spread can really be significant with the so-called penny stocks, and individuals who trade these stocks should be aware of how much it costs to do business.

Computing the Height of Objects

This topic could have been placed in other sections of the book, but we often want to know how high something is when we're traveling: a magnificent tree, a tall building, a distant mountain. In order to be able to compute this, it is necessary to be able to measure angles. Surveyors have instruments to do this, but it is possible to do so simply by drawing lines on a piece of paper that represent lines of sight to the top of the object in question. These angles can be measured with a protractor, a staple of high-school geometry courses. These days you don't have to buy a protractor; you can simply print one out (and rulers as well) from an assortment of online sites.

There are basically two ways to measure the height of a distant object, depending upon whether or not you can measure the distance directly from where you are to the base of an object.

When You Can Walk Up to the Base of an Object

This is basically a simple question in trigonometry.

Practical Math Problem

10.8 Height Measurement Using One Angle

What is the height of a tree that is 50 feet way from an observer, and the angle of elevation from the observer to the top of the tree is 62°?

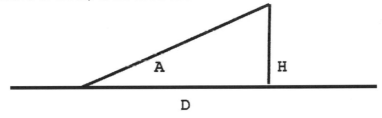

➤ D = distance from observer to object

➤ A = angle of elevation from observer to top of object

➤ H = height of object

$$H = D \tan A$$

The height of the tree is 94 feet.

This is called a one-angle height measurement because only one angle is needed to determine the height of the tree. However, if the tree is located on the other side of a river, or if the object is a distant mountain for which one cannot accurately determine how far it is to the point at ground level directly below the summit, you need to use two angles to determine the height.

Practical Math Problem

10.9 Height Measurement Using Two Angles

The angle of elevation to the top of a tree on the other side of a river is 40°. If you walk 100 feet in a straight line directly away from the tree, the new angle of elevation is 28°. How wide is the river? How tall is the tree? (As mentioned, this method is useful when one cannot directly measure the distance to the object, as in the preceding example.)

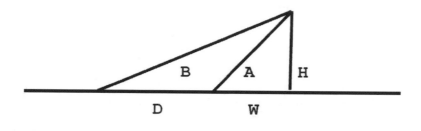

➤ A = larger angle of elevation

➤ B = smaller angle of elevation

➤ D = distance between angle measurements

➤ H = height of tree

➤ W = width of river

$$W = \frac{D \tan B}{\tan A - \tan B}$$

$$H = W \tan A = \frac{D \tan A \, \tan B}{\tan A - \tan B}$$

The width of the river is approximately 173 feet. The height of the tree is approximately 145 feet.

Distances by Triangulation

Angle measurement also provides a way to measure distances to objects. In problem 10.9, one can imagine that the straight line at the base is the shore of a body of water, and the top of the tree is an island located some distance away. The angles to the island are measured, and H now denotes the distance from the shore to the island.

In this case, the island is located on the same side of both places where the angles are measured. If the island is between the places where the angles are measured, a slightly different picture results.

Practical Math Problem

10.10 Distance by Triangulation

Two observers on a beach are separated by 3 miles. They see a boat at sea between them, and measure the angles between the line to the boat and the line between the two observers. One observer measures an angle of 35° and the other an angle of 61°. How far is the boat from the shore?

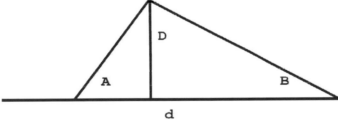

> ➤ A = angle from one observer to boat
> ➤ B = angle from second observer to boat
> ➤ d = distance between observers
> ➤ D = distance from observer to shore

$$D = \frac{d \tan A \tan B}{\tan A + \tan B}$$

The distance to the boat is 1.51 miles.

CHAPTER 11

 # To Your Good Health

---

In This Chapter

➤ Staying healthy by the numbers

➤ Getting the most for your dietary supplement dollar

➤ Administering drugs

Can I Skip This Chapter?

This chapter is intended for anyone who either monitors their own health or is responsible for doing so for others.

Staying Within Limits

One of the most important contributors to a happy life—to say nothing of a long one—is good health. Almost everybody knows that certain numbers are important. You want to keep your blood pressure down to help prevent a stroke, your cholesterol down to avoid a heart attack, and your glucose down to help prevent diabetes. Most people monitor these numbers on a fairly regular basis, and so it isn't really necessary to dwell on them here.

There are other numbers not so well known that are directly and indirectly connected with health. There are also a number of problems that arise in connection with health maintenance that have a relationship to practical math.

One of the most basic indices of the overall state of your health is how much you weigh for how tall you are. The body mass index is a good indicator of whether your weight is appropriate for your height.

The Body Mass Index (BMI)

The body mass index supplies a basic indicator to decide whether you are underweight, just about right, overweight, or seriously overweight.

Practical Math Problem

11.1 The Body Mass Index (BMI)

What is the Body Mass Index of a man who is 5 feet 9 inches tall and weighs 150 pounds?

➤ H = height in inches

➤ W = weight in pounds

➤ B = Body Mass Index

$$B = \frac{704.5\,W}{H^2}$$

His Body Mass Index is 22.2.

Since you're probably curious where you officially fit in, the ranges are as follows:

➤ 0–17: underweight

➤ 18–24: healthy

➤ 25–29: overweight

➤ 30–39: obese

➤ over 40: seriously obese

Admittedly, you probably have a good general idea of where you stand, but there is a tendency for us to engage in a bit of self-deception. Anorexics never think they are thin enough, and many obese and seriously obese people don't really see themselves as such. If you want to get an idea of what you need to weigh in order to bring your BMI within an acceptable range, the following alternative form should help you. After all, you're not going to be able to change your BMI by gaining or losing height.

Practical Math Problem

$$W = \frac{BH^2}{704.5}$$

This formula is used to answer questions such as "If I am 6 feet tall, what must I weigh so my BMI does not exceed 25?" The answer to this question is 184 pounds.

After taking an honest look at your BMI, you may have decided to go on a diet. After a little work with the above alternative forms formula, you've mapped out a target weight. How are you going to hit this target?

Practical Math Problem

11.2 A Schedule for a Diet

An overweight individual wants to go on a diet to go from his current weight of 220 pounds to 180 pounds within three months. How many pounds should he plan on trying to lose weekly?

➤ W = current weight

➤ T = target weight

➤ N = number of weeks for dieting

➤ P = pounds per week weight loss

$$P = \frac{W - T}{N}$$

Assuming there are thirteen weeks in the three months, the dieter should plan to lose about 3 pounds a week.

Many authorities agree that dieting sensibly and exercising is as effective as crash or fad diets and is more likely to result in permanent weight loss; weight that is lost quickly tends to return.

Problem 11.2 is actually fairly easy to compute, even without a calculator. Divide the number of pounds you need to lose by the number of weeks allotted to lose the weight, and that's the average number of pounds per week you need to lose.

You can apply problem 11.2 to a wide number of planned schedules. Do you need $150 for a Christmas present for that special someone and it's the middle of July? You've got about twenty weeks left to save the money, so you need to put aside $150 divided by 20, or $7.50 a week.

Are you a weight lifter who wants to bench press 50 more pounds in four months? That's sixteen weeks or so, so you just need to add an average of 3 pounds a week. If every two days you put an extra pound on the bar, you'll get there ahead of schedule.

You've Got to Have Heart

You get two each of arms, legs, hands, feet, ears and eyes at birth, but you're given only one heart, and it's important to take care of it. The heart is a muscle, and as you get older, it becomes more susceptible to excessive stress.

Practical Math Problem

11.3 Maximum Heart Rate

What is the maximum heart rate for a typical 55-year-old?

➤ A = age in years

➤ H = maximum heart rate in beats per minute

$$H = 212.3 - 0.812A$$

It's probably not a good idea to exceed 168 beats per minute.

This formula is an average of several of the leading estimates for this number and should be used as a guideline rather than an ironclad rule. Obviously, those in top physical shape can tolerate a higher maximum heart rate, while those in poor health should not expose themselves to unnecessary cardiac stress.

Good health is generally about guidelines rather than ironclad rules.

Soak Up the Sun—But Not Too Much!

If you really want to soak up the sun, you'd better put on some sun lotion, as melanomas can be anywhere from nasty to fatal. Fortunately, the manufacturers of sun block have stuck a number on the bottle—the SPF number, or sun protection factor, which enables you to determine how long you can stay out in the sun without damage—provided, of course, that you apply sun block.

Practical Math Problem

11.4 Time until Sunburn

What SPF factor sun block is needed to enable an individual to stay in the sun for 3 hours, if he normally experiences sunburn in 10 minutes without protection?

➤ B = time taken to experience sunburn without protection

➤ T = desired time to stay in direct sunlight

➤ S = SPF factor needed

$$S = \frac{T}{B}$$

He needs an SPF of 18 but should probably get sunblock with an SPF of 25 or 30 to stay on the safe side.

Most people have a fair idea, based on experience, of how long it takes them to experience sunburn. However, it isn't an experiment you want to try. It is generally true that the lighter the skin, the more quickly it burns—redheads and blonds tend to burn more quickly, while brunets and people with olive skin take longer to burn. Some conditions are altered by genetics, and people with Celtic heritage (generally Irish or Scottish) have increased risk of skin cancer and should be especially careful about staying out of the sun.

Blood Alcohol Level

Many of us drink, and most of us have imbibed too much on at least one occasion. The ideal thing to do is to have a designated driver, but that may not always be possible. The next best thing to do is to wait until it's safe to drive, but how long is that?

We can all drive with some fraction of alcohol in our blood, although some people do a better job than others when so impaired. However, the police have a test for measuring blood alcohol. In most states you can't drive if your blood alcohol level is above 0.08.

Practical Math Problem

11.5 Blood Alcohol Content

How long should a 150-pound female wait after consuming 6 ounces of 70 proof alcohol to make sure that her blood alcohol content is below 0.08?

➤ B = blood alcohol content level

➤ N = number of ounces consumed

➤ P = proof level of alcohol

➤ W = body weight in pounds

➤ r = 0.73 for men, 0.66 for women

➤ H = number of hours since drinking began to reach blood alcohol content level B

$$H = 66.67(\frac{.0257NP}{Wr} - B)$$

The formula tells us that she should wait about 1.93 hours, but to be on the safe side maybe 2.5 hours would be better.

There's another version of this formula that is also useful.

Practical Math Problem

$$N = \frac{Wr(B + .015H)}{.0257P}$$

This form is useful for answering questions such as "How many ounces of 86-proof whiskey can a 180-pound man drink in a 3-hour period and still remain under the 0.08 legal limit for blood alcohol content?" The answer to this question is 7.4 ounces.

It's probably too much to hope that a person in the process of drinking—possibly to excess—would stop to calculate her blood alcohol level or how long she should wait before driving if she's had too much to drink., so this formula presents calculations that should be performed in advance and carried around with you.

Saving Money on Dietary Supplements

The following situation arises with some frequency. Your doctor suggests that you make sure that you get 100 units of iron and 120 units of niacin daily. It would really be nice if someone manufactured a pill with 100 units of iron and 120 units of niacin, but that's probably not available. But your local health food store or drugstore offers a number of niacin-rich supplements with some iron, and iron-rich supplements with some niacin. What should you do?

We've actually seen a variation of this problem before. Problem 4.5 detailed a procedure for obtaining the maximum profit in a situation that called for selling two different types of trail mix made from two ingredients: fruit and nuts. This is the opposite side of the coin; we want to pay as little as possible rather than to make as much profit as possible.

Practical Math Problem

11.6 Saving on Dietary Supplements

A patient needs at least 100 units of iron and 120 units of niacin daily. Two supplements are available. The first supplement contains 5 units of iron and 20 units of niacin per ounce, and costs 20¢. The second supplement contains 10 units of iron and 5 units of niacin per ounce, and costs 30¢. What is the cheapest combination of the two supplements that will satisfy the daily nutritional needs, and how much of each supplement do you need?

Problems such as these are easier to read if the information is presented in table form. Here's one way to do it:

Data for 1 gram of each Supplement

	Iron (units)	Niacin (units)	Cost (cents)
Supplement 1	5	20	20
Supplement 2	10	5	30
Units required	100	120	

If we label iron and niacin as Items, we can construct a generic version of this table as follows. Of course, depending upon what is needed, we can't be sure whether we'll need to purchase grams or ounces of the supplements, but only the numbers that have been supplied above matter. If it took one gram (rather than one ounce) of each supplement to supply the quantities listed in the table, you'll just end up taking a lot less of each supplement but doing the same math.

Data for 1 ounce (or gram) of each Supplement

	Item 1	Item 2	Cost
Supplement 1	a_1	b_1	c_1
Supplement 2	a_2	b_2	c_2
Units required	A	B	

The solution to the problem will be as follows:

➤ X = number of units of supplement 1

➤ Y = number of units of supplement 2

➤ M = minimum cost

We now compute the minimum revenue M, which will be the smallest of the numbers P, Q, and R.

$$P = c_1 \max\left(\frac{A}{a_1}, \frac{B}{b_1}\right)$$

$$Q = c_2 \max\left(\frac{A}{a_2}, \frac{B}{b_2}\right)$$

$$R = \frac{c_1(Ab_2 - Ba_2) + c_2(Ba_1 - Ab_1)}{a_1 b_2 - a_2 b_1}$$

$$M = \min(P, Q, R)$$

You've now determined what the minimum revenue M will be. The next step is to determine X, the number of units (grams or ounces) of supplement 1 you must take, and Y, the number of units of supplement 2.

If M = P, you won't need to take any units of supplement 2.

$$X = \max\left(\frac{A}{a_1}, \frac{B}{b_1}\right) \quad Y = 0$$

If M = Q, you won't need to take any units of supplement 1.

$$X = 0 \quad Y = \max\left(\frac{A}{a_2}, \frac{B}{b_2}\right)$$

Finally, if M = R, you're going to have to take some of each.

$$X = \frac{Ab_2 - Ba_2}{a_1b_2 - a_2b_1} \quad Y = \frac{Ba_1 - Ab_1}{a_1b_2 - a_2b_1}$$

You should use 4 ounces of supplement 1 and 8 ounces of supplement 2 for a minimum cost of $3.20.

If in the unlikely chance that the minimum cost is obtained by using one of the numbers P, Q, or R and either of the associated values of X and Y are negative, eliminate that as a possible solution. Now look for the minimum cost from the remaining two numbers of P, Q, and R.

Just like problem 4.5, this is a lot of work, but this problem can appear in many different guises. Here's another variation of the same problem.

Practical Math Problem

11.6 Saving on Dietary Supplements in a Different Guise

A garage owner hires two mechanics, Al and Bob, on a daily basis. In a typical day, Al could repair 3 cars and 3 trucks. Al's daily salary is $190. Bob can repair 2 cars and 4 trucks per day and has a daily salary of $220. If there are 12 cars and 18 trucks that need repairs, how many days of each worker should the owner hire?

Even though it's a totally different environment, with Al and Bob playing the role of dietary supplements and cars and trucks substituting for iron and niacin, the idea is basically the same. The owner should hire Al for two days and Bob for three days at a total cost of $1,040.

Administering Drugs

Even though the problems in this section are nominally for people who are in the health care profession, the time may come when you are confronted with the problem of administering the correct dosage of a drug.

The Correct Dosage to Administer

Practical Math Problem

11.7 Correct Dosage

A patient needs 40 milligrams (mg) of a medication intramuscularly. The drug is available as 100 mg/5 ml. How many milliliters (ml) of the drug should be administered?

The notation "100 mg/5 ml" means that 100 mg of the drug are available in 5 ml of solution.

> ➤ D = amount of medication needed
>
> ➤ H = quantity of drug in vehicle V
>
> ➤ V = vehicle in which quantity H is available

The variables H and V could also be described as H units of medication available per V units of material conveying the drug for delivery.

> ➤ N = quantity of vehicle to be given to the patient

$$N = \frac{DV}{H}$$

The patient should be given 2 ml of the drug.

This is another problem that can be solved through ratio and proportion, or merely by having a reasonable comfort level with arithmetic. If 100 mg of the drug are available in 5 milliliters of the solution, that means that 20 mg of the drug are available in each milliliter of solution. So in order to obtain 40 mg of the drug, simply divide 40, the number of milligrams needed, by 20, the number of milligrams in each milliliter, to see how many milliliters need to be administered.

Practical Math Problem

11.8 Drug Dosage by Drips

500 ml of a drug are to be delivered intravenously over a 4-hour period. The tubing drip factor is 10 drips per milliliter (Note: *gtt* is a standard abbreviation for a single drip, should you ever see this notation). How many drips per minute should be administered?

➤ V = volume of drug in milliliters

➤ T = time for drug to be administered in hours

➤ D = tubing drip factor in drips per milliliter

➤ N = number of drips per minute

$$N = \frac{VD}{60\ T}$$

The drug should be administered at a drip rate of 20.8 per minute. Of course, it isn't generally possible to achieve that degree of accuracy, but fortunately that degree of accuracy is never needed; 20 drips per minute will suffice.

CHAPTER 12

 Recreation

In This Chapter

➤ The big three: baseball, football, and basketball

➤ Other games and activities

➤ Closing in on victory

Can I Skip This Chapter?

This chapter is devoted to the mathematics of sports and recreational activities. Some of it involves the statistical parameters that are used to evaluate success, and some involves the probabilities that are associated with sports and other activities. If you like sports, you don't want to skip this chapter.

Games and Practical Math

This chapter is mostly about sports. What's that got to do with practical math?

The last chapter, if you recall, concerned health. Granted, not all the material in the last chapter applied to you personally, but some of it undoubtedly did. Even the material that did not apply to you probably impressed you as having practical value for someone.

But what about sports; isn't that just basically frivolous fun and games? Possibly—but what of it? There are countless studies that show that fun and games keep us healthy, and you can't deny the practical aspect of that. However, there is practical value to the practical math in sports.

It is hoped that as you go through this book you become more acquainted with math. Admittedly, most people who use this book will use it for specific problems and situations that occur for which they need solutions, and that's why the book was written. However, the more math you are exposed to, even if it's just looking up a formula, the more comfortable you become with math. There have been countless studies that show this to be true. The more comfortable you are with math, the more you become aware of the types of situations in which math can help you.

There's a lot of math in sports. Many people who became interested in math got interested through sports, through the computation of batting averages and discussions of efficiency ratings for quarterbacks. Interestingly enough, the math that occurs in sports sometimes has applications to a much wider collection of problems; we'll see a couple of examples of that in this chapter.

Finally, we come to the other reason that this chapter is about practical math: what could be more practical than improving the lives of your children and your relationship with them?

Sports have created a bond between parents and children since the beginning of organized sports and probably even before, when the only sports were hunting and fishing. Sports are also a wonderful, and virtually painless, way to improve your child's ability with math. And what could be more practical than that?

Baseball, Football, and Basketball

These are the three biggest organized sports in the United States. Major League Baseball, the National Football League, and the National Basketball Association are multibillion-dollar industries. In addition, these games are played by children of all ages and by a large number of adults as well.

Baseball

Many children first become acquainted with averages, which are arguably the most important concept in mathematics, through their exposure to baseball. Although both football and basketball feature averages as an important way of assessing the performance of teams and individual players, baseball has the longest tradition of incorporation of averages as an important part of the game.

Practical Math Problem

12.1a Batting Averages

What is the batting average of a player who has had 350 official at bats and 87 hits?

> ➤ A = number of official at bats
>
> ➤ S = number of singles
>
> ➤ D = number of doubles
>
> ➤ T = number of triples
>
> ➤ H = number of home runs
>
> ➤ B = Batting Average = number of hits / number of official at-bats

$$B = \frac{S+D+T+H}{A}$$

The player is batting .249 (this is read as two forty-nine).

Unlike some so-called averages, the batting average really is an average in the mathematical sense of the term: it is the average number of hits per official at bat. Ever since the inception of the game, the batting average has been one of the key statistics used to assess the quality of a batter.

From a practical standpoint, computing batting averages is an excellent way to help a child become acquainted with equivalent fractions and their decimal values. From a baseball standpoint, the fraction 6/21 and 2/7 are equivalent because in each case the batter has gotten two hits for every official at bat. A child who is a baseball fan will quickly associate the decimal .286 with the fraction 2/7.

Practical Math Problem

12.1b Slugging Percentage

What is the slugging percentage of a player who had 27 home runs, 4 triples, 36 doubles, and 71 singles in 477 official at bats?

➤ S = Slugging percentage = total number of bases on hits / number of official at bats

$$S = \frac{S+2D+3T+4H}{A}$$

The player's slugging percentage is .551. The term *percentage* is a misnomer, for it is really a weighted average, with extra base hits being weighted according to the number of bases achieved.

Children as students have a slugging percentage as well. Counting an A as a home run (4 grade points), B as a triple (3 grade points), C as a double (2 grade points), D as a single (1 grade point), and F as ignominious outs, a student's grade point average (GPA) is the total number of bases (grade points) divided by the number of official courses completed.

Actually, there is a slight difference. Grades in courses such as music appreciation or physical education may not be counted in the official compilation of a student's GPA. In college, some courses (such as science courses that require lab work) are more heavily weighted. The GPA in this case is computed by assuming that each unit has a grade point attached to it: an A in a 3-unit course gets 12 grade points (3 x 4). The GPA in college is the total number of grade points divided by the total number of units in completed courses.

Practical Math Problem

12.1c Earned Run Average

What is the earned run average of a pitcher who allowed 23 earned runs in 85 2/3 innings pitched?

Earned run average = 9 × number of earned runs allowed / number of innings pitched

The earned run average is the time-honored way of assessing the performance of a pitcher. The pitcher in 12.1c had an earned run average of 2.42.

Earned run averages give a child an excellent way to become familiar with fractional computations and ratio and proportion. If a pitcher allows one earned run in two innings, his earned run average can be computed by the above formula. The proportion below also enables us to make this computation.

1:2 = ERA:9

Football

It wouldn't be very interesting to present formulas for football similar to ones for baseball (such as the yards per carry for a running back or a quarterback's completion percentage), so let's look at the measure of a quarterback's ability: the quarterback efficiency rating.

Practical Math Problem

12.2 NCAA Quarterback Efficiency Rating

A college quarterback threw 25 passes in a game, completing 14 of them for 205 yards and 1 touchdown. Two passes were intercepted. What is his efficiency rating?

> ➤ A = number of passes thrown
> ➤ C = number of completions
> ➤ I = number of interceptions
> ➤ T = number of touchdowns
> ➤ Y = number of yards
> ➤ E = efficiency rating

$$E = \frac{100C + 330T + 8.4Y - 200I}{A}$$

The quarterback's efficiency rating is 122.

The National Football League (NFL) has a similar formula, but it is much more complicated.

This is an example of the use of mathematical models for assessment purposes. Another well-known example of the use of mathematical models for assessment purposes is the concept of moneyball, which attempts to value players by using their individual statistics and the salaries they are being paid.

Basketball

Late in a game, a player is often fouled to make him shoot free throws. The following situation brings up some very interesting additional points.

Practical Math Problem

12.3 Probability of Making Free Throws

A player who makes 80% of his free throws is shooting 3 free throws. What is the probability that he will make exactly 2 free throws?

➤ p = probability of a successful free throw (expressed as a decimal)

➤ N = number of free throws attempted

➤ k = number of successful free throws

➤ P = probability of making exactly k of N free throws (expressed as a decimal)

$$P = \frac{p^k(1-p)^{N-k}}{k!(N-k)!}$$

The probability of making exactly 2 of 3 free throws is 38.4%.

In describing probabilities, rather than computing with them, it is generally more informative to express them as percentages rather than as decimals. However, in many formulas for probabilities, the probabilities themselves are expressed as decimals rather than percentages (e.g., a probability of 0.37 rather than a probability of 37%) because the formulas become extremely unwieldy if percentages are used rather than decimals.

Probability plays an important role in sports. Successful sports strategy is often based on knowing the probabilities associated with certain situations. Making decisions on the basis of these probabilities is known as playing the percentages. Managers or coaches who disregard the percentages and win are often described as brilliant, but managers who disregard the percentages and lose often find themselves looking for another job.

There are some assumptions that are involved in the above computation. The most important one is that the probability of making each free throw is independent of all other factors; it is 80% no matter whether it is early in the first quarter or the game is on the line in the last quarter. Independent probabilities also are not affected by previous results. It does not matter whether the shooter has missed or made his last five free throws; the probability of making his next free throw is 80%.

The formula associated with this problem is known as the Bernoulli trial formula to honor the mathematician who did the most to investigate the problem, although it is sometimes called the binomial trials formula. It generalizes from free throws to independent trials and is important not only in the determining probabilities, but also plays a fundamental role in statistics.

If we saw a player shoot twenty free throws and make only five of them, we would probably conclude he was a lousy free throw shooter. Admittedly, an 80% free throw shooter might very well make only ten of twenty free throws, but the probability of his doing that is only about 1 in 500. Similarly, if a drug manufacturer claims that a particular drug is effective in 80% of the cases, but it only works on ten of twenty people in a test, we would be very skeptical that the drug really is effective in 80% of the cases.

What started out as a basketball problem morphed into a much more general one. That's why getting children interested in the mathematics associated with sports is a very practical thing to do.

Other Sports

The Big Three aren't the only games in sports. NASCAR and drag racing are gaining in popularity.

Practical Math Problem

12.4 Final Velocity of a Dragster

From a standing start, a drag racer covers a ¼ mile in 11 seconds. Assuming that his acceleration is constant, what is his velocity at the end of the ¼ mile?

➤ D = distance covered in miles

➤ T = time taken in seconds

➤ V = terminal velocity in miles per hour

$$V = \frac{7200D}{T}$$

The car is going 164 miles per hour when it hits the finish line.

This is actually a problem in physics, and it is just another example of how different areas of knowledge impact sports.

Probability of Winning by 2 Points

Many games, such as tennis, ping-pong, and volleyball, have a fixed number of points, but you must win by 2 points.

In ping-pong, the first player who gets to 11 wins, but he must win by 2 points. So when the score is 10 to 10, if one player wins the next 2 points, he wins the game. If the players each win 1 of the next 2 points, the score is 11 to 11, and the pattern repeats. Either one player wins the next 2 points to win the game, or the players each win 1 of the next 2 points, and the game continues from 12 to 12 according to the same rules.

Practical Math Problem

12.5 Probability of Winning by 2 Points

You have reached a moment in a game where the first person to score 2 more points than his opponent wins (such as 10 to 10 in ping-pong or deuce at tennis). If your probability of winning each point is 60%, what is your probability of winning the game?

➤ p = probability of winning a point (expressed as a decimal)

➤ P = probability of winning a game (expressed as a decimal)

$$P = \frac{p^2}{1 - 2p(1 - p)}$$

Your probability of winning the game is 69.23%.

Computing Your Golf Handicap

There used to be a time when computing your golf handicap was fairly easy. You'd play a number of rounds on your local course, take a percentage of the average difference between your score and par, and presto—there's your handicap.

Of course, this didn't take into account whether your local course was the notorious Bethpage Black on which even the best professionals find it difficult to break par, or a course with relatively short and straight holes, few sand traps, and no water. The United States Golf Association (USGA) decided to come up with a method to correct for inequities, so here's how it's done.

Practical Math Problem

12.6 Computing Your Golf Handicap

You recently took a golfing vacation to Farmingdale, Long Island, where you played five rounds of golf on the notorious Bethpage Black, with a course rating of 75.4 and a slope factor of 144. Your scores were 97 (you got lucky), 108 (lots of rough), 99 (actually it was 100, but your club really didn't touch the ball that time, so you didn't count that stroke), 102, and 106. What is your handicap as determined by these five rounds?

You need to compute your handicap differential for each of the five rounds. In order to do this, you need your score and two numbers from the golf course, which is published by the USGA.

➤ Y = your score (which you always report accurately, because that's the kind of person you are)

➤ C = course rating (to distinguish Bethpage Black from creampuff courses)

➤ S = slope rating (a number that is used to correct for the fact that better golfers do much better on tough courses than poor golfers, but only somewhat better than poor golfers on easier courses)

From these two numbers you compute your handicap differential H for the round you just played.

$$H = \frac{113(Y - C)}{S}$$

The handicap differentials for your five scores were 16.95, 25.58, 18.52, 20.87, and 24.01.

Your handicap can be computed from as few as five rounds or as many as twenty. If you have played more than twenty, you use the most recent twenty scores. Depending upon how many rounds you play, you select a few of your lowest scores. There's a sliding scale (consult the USGA website). If you use five or six rounds to determine your handicap, you use your lowest handicap differential; if you use twenty rounds, you use your ten lowest handicap differentials.

Take 96% of the selected lowest handicap differentials, round down to the nearest tenth, and that's your handicap. In your case, you would use only the 16.95 because you only had five handicap differentials available, and 96% of that is 16.27, so your handicap is 16.2.

Bird-Watching

In order to relax, you decide that instead of trying to make birdies at Bethpage Black, you'll watch some in the local woods. But you can't help but be competitive, so you go to the local Audubon Society, pick up its field guide to American birds, and wonder how many birds you'll have to watch until you can enter all the local species on your life list.

This is a question that is difficult to answer if there are a few rare birds in a particular environment. But if you're dealing with large populations and any one type of bird is as likely as any other, there's an answer to this question.

Practical Math Problem

12.7 Observing All Species of Birds

A local bird-watching venue has 15 different species of birds, and the probability of seeing a bird of any 1 species is approximately the same as seeing a bird of any other species. On average, how many birds will you have to observe before you've seen 1 of each species?

➤ N = number of different species
➤ T = average number of observations before seeing one of each species

$$T = N(\frac{1}{1} + \frac{1}{2} + \cdots + \frac{1}{N})$$

You can expect to observe about 50 birds (the actual average is 49.77) before seeing one of each type.

Of course, this is an average. If you had one hundred different bird-watchers and asked each how many birds he or she had observed before seeing one of each species, you might find some who had seen as few as thirty, and some who had to see almost one hundred, but the average of the numbers from the one hundred bird-watchers would be about fifty.

Suppose that you decide to extend your vacation and go to an area in which there is a certain type of bird that is very rare. On average you might only see a single bird of this type in a full day of bird-watching. How likely is it that you'll see several in the same day?

Practical Math Problem

12.8 Observing Several Rare Birds

What is the probability of seeing 3 birds of a particular rare species in the same day if on an average day, you will only see 1?

> ➤ N = number of expected occurrences of the given event in the interval in question (in 12.8, the number of expected occurrences is 1 in one day, which is the interval in question)
>
> ➤ k = number of occurrences for which the probability is sought (3 in 12.8)
>
> ➤ P = probability of k occurrences in the interval in question

$$P = \frac{N^k e^{-N}}{k!}$$

You would only see this happen about 6% of the time. In other words, in one hundred days of bird-watching, you'd only see three birds of this species on six of those days. That's also the average. If you were to ask one hundred bird-watchers each to go bird-watching for one hundred days and count the number of days on which they saw three birds of this species; the average for the one hundred bird-watchers would be about six.

This is another famous problem from probability that has substantial application outside bird-watching, just as the Bernoulli trial formula has substantial application beyond shooting free throws. This particular formula is known as Poisson law of small numbers. It applies whenever one wishes to calculate the probability of a certain number of random events happening during a given interval for which the expected number of events can be calculated. It applies to problems such as the probability of a number of auto accidents in a week when the average daily number of accidents is known, or the probability of a specific number of auto accidents in a certain stretch of road whose length is known, given that accidents occur randomly along the road and the expected number of accidents per mile is known.

Knowing What Is Needed to Clinch Victory

The emphasis in competitive sports is on winning, and there's some interesting practical math that is used when victory is in sight.

Magic Numbers

The magic number is generally computed for teams in a league occupying first place toward the end of the season and is the total number of its wins plus the losses of the second-place team that will ensure that the team currently occupying first place ends up in first place.

Practical Math Problem

12.9 Magic Number for a First-Place Team

On February 22, 2011, the San Antonio Spurs were in first place in the Southwest Division of the NBA's Western Conference, with a record of 46 to 10 (46 wins, 10 losses). In second place were the Dallas Mavericks, with a record of 40 to 16. The NBA season is 82 games long. What was the magic number for the Spurs at that time?

➤ S = length of season

➤ W = number of wins of first-place team

➤ L = number of losses of second-place team

➤ M = magic number for first-place team

$$M = S - W - L + 1$$

The magic number for the Spurs is 21.

There is a variation of this problem that applies to computing needed win percentages.

Practical Math Problem

12.10 Win Percentage Needed

A baseball team has played 100 of its 162 games and has won 58 of them. All 35 of its remaining games are at home, where it wins 65% of its games. What percentage of the road games must it win in order to have an overall winning percentage of at least 60%?

➤ N = number of games in season

➤ n = number of games played to date

➤ H = number of home games remaining

➤ w = number of games won to date

➤ h = home winning percentage

➤ p = desired overall winning percentage

➤ r = needed road winning percentage

$$r = \frac{pN - 100w - hH}{N - n - H}$$

It needs to win 60.93% of its road games, which translates to winning seventeen of its remaining twenty-seven road games.

There is a version of this program that becomes prominent during every major election.

Practical Math Problem

12.11 Percentage of Uncounted Ballots

With 80% of the vote counted, a candidate has 55% of the vote. What percentage of the remaining ballots must vote for him in order to ensure he gets a majority of the votes?

> ➤ p = percentage of ballots counted
> ➤ v = percentage of counted ballots for candidate
> ➤ P = percentage needed of remaining ballots

$$P > \frac{5000 - pv}{100 - p}$$

He needs to win more than 30% of the uncounted ballots.

Life really does mirror sports, and the practical math of different aspects of life mirrors the practical math of sports as well.

CHAPTER 13

Understanding Gaming

In This Chapter

➤ Expected value

➤ The house percentage

➤ Hedging your bet

Can I Skip This Chapter?

Although this chapter is aimed primarily at people who want to understand the mathematics of gambling, the key concept of expected value underlies many business decisions. It is also useful to learn how to hedge your bets to effectively manage risk.

Gaming: It's More than Gambling

Gaming is the recently invented term for what used to be called gambling. It was felt that gambling had too many negative connotations, especially in the late 80s and early 90s as Las Vegas shifted its focus from attracting high rollers to attracting families. This was a remarkably successful business decision, and the practical math that underlies gaming is very similar to the practical math that underlies a large number of business decisions.

Expected Value

Expected value is one of the most useful concepts of practical mathematics. Many decisions are like bets on sporting events: they are made in an atmosphere of uncertainty. When a business launches a new product, it's a gamble. When an employee asks his boss for a raise, it's a gamble. When a guy asks a girl for a date, it's a gamble. The aspects common to all situations is that the venture will succeed some of the time. There is a reward associated with success, and a penalty associated with failure.

In the case of the bet on the sporting event or the launch of a new product, the rewards and penalties are measured in dollars. These are the situations that are easiest to analyze because both rewards and penalties have convenient numerical values associated with them.

Practical Math Problem

13.1 Expected Value of a Bet

You have bet $20 at odds of 5-2 (meaning that if you win, for every $2 you risked, you will win $5) on a team that you feel has a 30% chance of winning. What is the expected value of your bet?

➤ p = probability of winning (expressed as a percentage)

➤ W = amount won on a winning bet

➤ L = amount lost on a losing bet

➤ E = expected value of bet

$$E = .01(pW - (100 - p)L)$$

The expected value of this bet is $1. This means that if your assessment of the probability of winning is accurate, and this situation comes up a number of times, your average win would be $1.

You can see where this number comes from. If your assessment of the probability of winning is accurate and you make ten bets, you can expect to win three times at $50 for each win. You can also expect to lose seven times at $20 for each loss. That's a net win of $10 in ten bets, or $1 a bet, the expected value.

Another way of assessing the expected value is as a percentage of the amount risked. You are risking $20 and your expected value is $1, which is 5% of $20, so your percentage expectation is 5%.

We've actually seen expected value before very early in the book. In problem 4.2, we looked at the percentage that a claim would need to be filed on an insurance policy in order to justify buying the more expensive, low-deductible policy. This is actually a version of an expected value calculation: we were computing the percentage at which the expected value for the higher premium (lower deductible) policy exceeded the expected value for the policy with the higher deductible.

Figuring the House Percentage

There are many different ways to bet on a sports event, but certainly one of the most popular is to bet at even money against the spread. Of course, you aren't really betting at even money, where you bet $10 and if you win, you win $10, and if you lose, you lose $10. Typically, you bet at 11-10: if you win, you win $10, but if you lose, you lose $11.

The way that this arrangement is normally described is by writing it down as -110 +100. This is a typical line for betting against a spread. If one bets what is called the money line, giving odds by betting on a favorite or taking odds by betting on the underdog, the line might be listed as something like -250 +210. This means that if you bet on the favorite, you have to bet $250 to win $100 (or $50 to win $20, because you are giving odds of 5-2 when you bet $250 to win $100), but if you bet on the underdog, you risk $100 to win $210. The institution making the line—what used to be called the book—makes its money from this disparity between the $250 you must risk on the favorite to win $100, and the $210 you will win if you risk $100 on the underdog.

We've encountered a version of this before in two different environments: when we were exchanging money, and when we were buying or selling stock on the bid and asked. Once again, the ideas one encounters in the sports world appear in the business world as well.

Just as we need to know how much we are losing to the discrepancy in exchange rates, or the bid-asked in stock transactions, we need to know how much we are losing to a betting line.

Practical Math Problem

13.2 House Percentage: Two-Sided Bet

We are calling this a two-sided bet simply because you can take either side of the line. Yes, there is a different type of line involving several options; we'll discuss that shortly.

What is the house percentage on a game in which the line is -250 + 210?

➤ F = dollars bet on favorite to win $1 ($2.50 in 13.2)

➤ U = dollars won on underdog for a $1 bet ($2.10 in 13.2)

➤ H = house percentage

$$H = \frac{100(F - U)}{F(U + 2) + 1}$$

The house percentage is 3.56 percent. The practical consequence of this number is that over the long run, if you were to simply pick favorite or underdog by flipping a coin or some other chance mechanism, you would lose 3.56% of your money.

Frankly, it's hard to understand why anyone would want to bet on horse races by going to the track, where the house percentage is sometimes in excess of 20%.

Some people bet on sports events for the entertainment value. It makes a game more exciting if there is something in it for you if your team wins. Some people bet sports the same way others play the stock market in the hopes of being sufficiently successful to make a living. The big difference is that betting the stock market has historically been a winning bet: you would have beaten inflation over the history of the stock market simply by picking stocks at random.

Can You Win Consistently Betting Sports?

Some people make a living—a very good living—betting sports. You cannot make a living playing games such as craps or roulette because the house has a built-in advantage that in the long run cannot be overcome.

However, it is possible to win consistently betting sports because although you are betting against the house as an agent, the house tries to set the line in an attempt to induce an equal amount of betting on both sides. That way, the house is guaranteed to win the house percentage.

Suppose that the line on the Dallas Cowboys-New York Giants football game is Dallas: 3 at the standard 11 to 10 odds. This means that if you bet on Dallas, Dallas must win by more than 3 points in order for you to win $10. If Dallas wins by exactly 3 points, no money changes hands (this is called a push), and if Dallas either wins by less than 3 points or loses, you lose $11. Similarly, if you bet on New York and the Giants either win or lose by less than 3 points, you win $10; if they lose by exactly 3 points, no money changes hands; and if they lose by 4 or more points, you lose $11.

Let's say $1,000,000 is bet on each side. Assuming the game does not end with Dallas winning by exactly 3 points, the house pays $1,000,000 to the winning bettors and collects $1,100,000 from the losing bettors, a profit to the house of $100,000.

However, if $1,500,000 is bet on one team and $500,000 on the other, the house still has the same mathematical advantage conferred by the 11-10 odds, but it has incurred substantial risk. If the winning side is the one that has bet $500,000, the house collects $1,650,000 from the losers and pays $500,000 to the winners; a profit to the house of $1,150,000. On the other hand, if the winning side is the one that has bet $1,500,000, the house collects $550,000 from the losers but must pay $1,500,000 to the winners; a loss to the house of $950,000. The house would far rather see a steady source of $100,000 every game than either win $1,150,000 or lose $950,000, and so the house tries to set the line to attract equal betting on each side.

Dallas could actually win 75% of the games in this instance and the house wouldn't care; the same amount of money was bet on both sides, so the house still emerges with a profit. A successful sports bettor makes money by being able to judge when the line, which is set to attract an equal amount of money on both sides, is wrong and one side or the other has a much better chance than 50% of winning.

In order to win at sports betting, though, you have to be able to do better than pick more than 50% winners on bets that are 11-10 against you. You have to overcome the house percentage.

Practical Math Problem

13.3a Overcoming the House Percentage

You are betting $1 to win $1.20. What percentage of bets must you win in order to break even?

➤ W = dollars won if $1 lost on a losing bet ($1.20 in 13.3)

➤ P = win percentage needed to break even

$$P = \frac{100}{1 + W}$$

You need to win 45.45% of your bets. This may not appear to be such a difficult proposition, but you should examine this in the context of what randomly selecting a bet would do in the face of a given house percentage.

Practical Math Problem

13.3b Win Percentage of a Random Bet

Suppose that you are betting $1 to win $1.20 in a situation where the house percentage is 4%. What is the win probability of a similarly placed random bet in which someone randomly chose to bet $1 to win $1.20 where the house percentage is 4%?

➤ W = dollars won if $1 lost on a losing bet ($1.20 in 13.3)

➤ H = house percentage

➤ P = win percentage needed to break even

$$P = \frac{100 - H}{1 + W}$$

The win percentage of a randomly placed bet is 43.64%.

The ratio of the break-even percentage to the win percentage of a randomly placed bet is 100:100-H, which shows how much better you have to be than random to be successful. If the house percentage is 4%, you only need to be 4.17% better than random to be a winner. On the other hand, if the house percentage is 20%, you need to be 25% better than random to be a winner.

Bets with Many Contestants

There are many situations in which the bettor is offered not just a choice between one side or the other, but several different possible choices at different odds. Horse races generally have more than one horse in the field. Golf and tennis tournaments always begin with many entrants. At the start of the football (or baseball) season, you can bet which team will win the Super Bowl (or World Series).

Practical Math Problem

13.4 House Percentage: Many Contestants

In a 3-horse race, the odds on the 3 horses are 6-5, 2-1, and 5-2. What is the house percentage on this race?

➤ R_1 = win amount for $1 bet on first contestant (6-5 is 1.2-1, a payoff of $1.20)

➤ R_2 = win amount for $1 bet on second contestant

… (You'll see these dots with some frequency. They indicate that there are terms missing. For instance, in this case there are missing terms between the win amount for the second contestant and the win amount for the last contestant.)

➤ R_n = win amount for $1 bet on last contestant

➤ H = house percentage

We first need to perform an intermediate computation, in order to simplify the form of the final result.

$$X = \frac{1}{R_1 + 1} + \frac{1}{R_2 + 1} + \cdots + \frac{1}{R_n + 1}$$

We can now compute the house percentage.

$$H = \frac{100(X - 1)}{X}$$

The house percentage is 6.85 percent.

Parlay Betting

A parlay bet is a bet on which several teams have to win in order for you to collect. The standard two-team parlay offers odds of 13-5 if both teams win. Let's assume that Dallas is favored by 3½ points over New York, and Green Bay is favored by 5½ points over Chicago, and you make a parlay bet on both Dallas and Green Bay. Half-point lines are used here in order to make sure that both games end in a win or a loss for the bettor, as the rules on how to resolve parlay bets are unnecessarily complicated when one or both games end up with no decision. If you have bet $10 on the parlay and both Dallas and Green Bay win, you win $26; otherwise you lose your $10.

Practical Math Problem

13.5 House Percentage on a Parlay

What is the house percentage on a 2-team parlay at odds of 13-5?

➤ W = dollars won if $1 lost on a losing parlay bet ($2.60 in 13-5)

➤ N = number of teams in parlay

➤ H = house percentage

$$H = \frac{100(2^N - 1 - W)}{2^N}$$

The house percentage on a 13-5 parlay is 10%.

Parlay bets can usually be made on any number of teams, but the larger the number of teams, the higher the house advantage. For instance, a typical three-team parlay pays 6-1 on a winning bet, which is a 12.5% house advantage.

Because the house percentage on parlay bets is so high—much higher than the house percentage on individual bets—parlay bets are often described as sucker bets.

Hedging Your Bet

Some events feature the opportunity to make another bet after the event has begun. If you bet on a team at the outset of the season to win the Super Bowl, you can still make bets on which team will win the Super Bowl as long as the season is still in progress. If you make a bet on a golfer to win a tournament and that golfer is still in the field on the last day of the tournament, you can make bets then. Some events even feature the opportunity to make bets in play; odds are continuously posted as the match develops.

This offers the opportunity to hedge. Similar situations also occur in the financial world. Hedging enables a bettor to take advantage of a winning position or to minimize downside risk from a losing one.

Practical Math Problem

13.6 Changing Horses in Midstream

A bettor has bet $300 to win on the underdog in a tennis match and received 3-1 odds. The underdog wins the first set, goes up a break in the second, and the odds change so that the bettor can now get 2-1 on the original favorite. How much should he bet at those odds to be sure of winning the same amount no matter who wins, and how much will that amount be?

➤ r = initial odds to 1 of bettor's bet (r = 3 in 13.6)

➤ N = initial amount of bet, receiving odds of r to 1

➤ q = later odds to 1 on opponent (q = 2 in 13.6)

➤ B = amount bettor should bet on opponent

➤ W = guaranteed winnings

$$B = N\frac{r+1}{q+1}$$

$$W = N\frac{rq-1}{q+1}$$

The bettor should bet $400 at odds of 2-1 in order to receive $500 no matter who wins.

Although the example given is one in which the bettor can ensure a win, the bettor can ensure an equal result by betting the amount B at q to 1. If rq > 1, this result will be a win, and if rq < 1, this result will be a loss. For instance, if the odds on the favorite have improved to 0.25-1 (a bet of $1 will risk $1 to win 25¢), by betting $960 on the favorite to win, the bettor can ensure a loss of $60 no matter who wins.

The same type of arrangement can be made even if there are several different opponents, all of whom receive different odds.

Practical Math Problem

13.7 Hedging with Multiple Contestants

A bettor has bet $200 on the favorite in a four-person event at odds of 2-1. He later can obtain odds of 7-1, 11-1, and 23-1 on the other three entrants. How much should he bet on each to assure himself of the same return no matter who wins?

➤ R = odds to 1 on bet already made (R = 2 in 13.7)

➤ X = amount of bet already made

➤ R_1 = odds to 1 on first team (or player)

...

➤ R_N = odds to 1 on last team (or player, N = 3 in the example)

➤ R = odds to 1 on bettor's money

➤ X_1 = amount to bet on first team

...

➤ X_N = amount to bet on last team

➤ In 13.7, R_1 = 7, R_2 = 11, and R_3 = 23

$$X_1 = X\frac{R + 1}{R_1 + 1}$$

$$\cdots$$

$$X_N = X\frac{R + 1}{R_N + 1}$$

The bettor should bet $75 at 7-1, $50 at 11-1, and $25 at 23-1. No matter which team wins, the bettor will win $250.

A similar situation can arise in which there is an event with multiple contestants. You want to bet on a few of them in such a way that no matter which one wins, you receive the same amount. For instance, you might decide that the three teams most likely to win the 2012 Super Bowl are Green Bay, Pittsburgh, and Indianapolis. Before the season begins, they are all likely to be posted with different odds.

Practical Math Problem

13.8 Multiple Bets with Equal Returns

A bettor wishes to bet on three teams in such a way that he will receive the same payoff no matter which of the three teams wins. The odds against the three teams are 7-1, 11-1, and 23-1. If he has $120 to bet, how much should he bet on each team, and what odds is he receiving on his money?

> ➤ R_1 = odds to 1 on first team (or player, R1 = 7 in 13.8, etc.)

> ➤ R_2 = odds to 1 on second team (or player)

...

> ➤ R_N = odds to 1 on last team (or player, N = 3 in the example)

> ➤ R = odds to 1 on bettor's money

> ➤ T = total amount available for betting (T = $120 in 13.8)

> ➤ X_1 = amount to bet on first team

...

> ➤ X_N = amount to bet on last team

$$Q_1 = \frac{1}{R_1 + 1}$$

$$\cdots$$

$$Q_N = \frac{1}{R_N + 1}$$

$$Q = Q_1 + \cdots + Q_N$$

$$X_1 = T\frac{Q_1}{Q}$$

$$\cdots$$

$$X_N = T\frac{Q_N}{Q}$$

$$R = \frac{1 - Q}{Q}$$

The bettor should bet $60 at 7-1, $40 at 11-1, and $20 at 23-1. He will receive odds of 3-1 on his money. If any of his three teams wins, he will make $360; if they all lose, he will lose $120.

Although the calculations are a little different, they are essentially the same type of calculations that financial experts managing funds make. Reward, risk, and return are the same concepts, whether one is investing in stocks or betting on sports events.

Gambler's Ruin

Even if you have positive expectations, you can lose if your opponent is better capitalized. For instance, if you have an 80% chance of winning a bet at which you are receiving 3-1 odds, your percentage expectation is an astounding 220%. But if you have only enough money for one bet, you could still lose.

Even if you had enough money for five bets—or ten bets—you could still be tremendously

Practical Math Problem

13.9 Gambler's Ruin

Assume that you have a stake of $50 and will continually make bets of $1 on a proposition that you feel you have a probability of .51 of winning. What is the probability that you will increase your stake to $125 before you go broke?

unlucky and lose. The situation in 13.9 illustrates this.

➤ S = your stake in units (50 in 13.9)

➤ N = stake goal in units (125 in 13.9)

➤ p = probability of winning a single bet (expressed as a decimal, not a percentage)

➤ P = probability of stake reaching goal before going broke

$$P = \frac{S}{N}$$

if p = ½

$$r = \frac{1-p}{p}.$$

if p ≠ ½,

In this case, we have to use a different formula than the one we used when p = ½.

$$P = \frac{1 - r^S}{1 - r^N}$$

You have approximately an 87% chance of having your stake reach $125 before you go broke.

Gambler's ruin is a problem not just faced by gamblers. Insurance companies must consider the possibility of paying a large number of claims; movie-production companies must consider the possibility of having a string of flops even though they have a track record of overall success.

CHAPTER 14

 In the Workshop

In This Chapter

> ➤ Gears and belts
> ➤ Drills and cutters
> ➤ Geometry and physical characteristics of cylindrical rods

Can I Skip This Chapter?

This is the first of several chapters on the practical math associated with using machines to do physical work, and it is designed to help those who spend a lot of time working with these machines. For those interested in going further than this chapter, *Machinery's Handbook* by Erik Oberg et al., which is now in its 28th edition, is an indispensable reference in this area.

The math in this chapter centers around the basic machines that one finds in a workshop: lathes, cutters, drill presses, and the like. There is also a section that centers on various properties of cylindrical rods, because cylindrical rods play an important part in many machines.

Gears and Belts

Many machines make use of both gears and belts. Gears are particularly important; they are central to autos and bicycles, and are one of the primary ways of changing both the speed and direction of rotary motion.

Practical Math Problem

14.1 Speed of a Driven Gear

A gear with 16 teeth is rotating at 120 rpm. It is driving a gear with 24 teeth. At what speed is the driven gear rotating?

> ➤ T = number of teeth in driving gear
> ➤ R = speed (in rpm) of driving gear
> ➤ t = number of teeth in driven gear
> ➤ r = speed (in rpm) of driven gear

$$r = \frac{T}{t}R$$

The driven gear is rotating at 80 rpm.

Alternative Forms

$$t = \frac{R}{r}T$$

This formula answers questions such as "How many teeth should there be on a gear that meshes with a gear having 60 teeth, if the speed of the 60-toothed gear is 400 rpm and it is desired that the meshed gear should rotate at 280 rpm?" The answer to this question is 42.

The two formulas above are the most basic ones for gears because they address one of the fundamental purposes of gears: changing the speed of rotation. However, each of the many types of gears (spur, worm, helical, etc.) has its own internal and elegant geometry. This generally defines what is needed to design gears rather than to use them.

Gears are not the only device available for changing the speed of rotation. Wheels connected by a belt also serve this purpose.

Practical Math Problem

14.2 Speed of a Wheel Connected by a Belt

A wheel with a 2-foot diameter is rotating at 20 rpm. It is linked by a belt to a wheel with an 8-inch diameter. At what speed is the driven wheel rotating if there is no slippage?

➤ D = diameter of driving wheel

➤ R = speed (in rpm) of driving wheel

➤ d = diameter of driven wheel

➤ r = speed (in rpm) of driven wheel

$$r = \frac{D}{d}R$$

The driven wheel rotates at 60 rpm.

Alternative Forms

$$d = \frac{R}{r}D$$

This formula answers questions such as "What should the diameter be of a driven wheel connected by a belt rotating without slippage to a wheel with an 8-inch diameter that is rotating at 120 rpm if it is desired that the driven wheel rotate at 150 rpm? The answer to this question is that the driven wheel should have a diameter of 6.4 inches.

Drills

Drills, as everyone knows, create holes. The critical questions involving drills are how wide do you want to drill the hole, how deep must you feed the drill, and how fast must the drill rotate?

Practical Math Problem

14.3 Drill Feed

The drill feed is the distance that the drill advances in a single revolution. It is usually measured in inches per revolution.

What is the feed of a drill that rotates at 300 rpm and drills a hole 2 inches deep in 20 seconds?

➤ R = rotation rate in rpm

➤ D = depth of hole

➤ T = time to drill hole (in minutes)

➤ F = drill feed

$$F = \frac{D}{RT}$$

The drill feed is 0.02 inches per revolution.

The other important relation for drill parameters is how the hole depth relates to the hole diameter.

Practical Math Problem

14.4 Required Drill Depth

How deep should you feed a 74° countersink to drill a hole 0.65 inches in diameter?

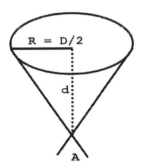

➤ A = angle of countersink

➤ D = hole diameter

➤ d = required depth

$$d = \frac{D}{2 \tan \frac{A}{2}}$$

The required depth is 0.43 inches.

Alternative Forms

$$D = 2d \tan \frac{A}{2}$$

This formula answers questions such as "What is the diameter of a hole when a drill with a countersink angle of 65° is sunk to a depth of 0.75 inches? The answer to this question is 0.96 inches.

Cutting Mills and Lathes

Like drills, cutting mills and lathes are tools commonly found in workshops for working with materials such as wood and metals.

Practical Math Problem

14.5 Cutting Speed of a Mill or Lathe

How many revolutions per minute must a cutting mill with a 2-inch diameter make in order to cut machine steel at 80 surface feet per minute?

➤ D = diameter in inches

➤ S = desired surface feet per minute

➤ R = revolutions per minute

$$R = \frac{3.82S}{D}$$

The cutting mill must rotate at 153 rpm.

Alternative Forms

$$= 0.261RD$$

This formula answers questions such as "What is the surface feet per minute cut by a 3-inch diameter drill revolving at 180 rpm?" The answer to this question is 141 surface feet per minute.

Cylindrical Rods

It's hard to imagine machines of consequence that do not include cylindrical rods somewhere in their construction. As soon as either wheels or gears appear, they are mounted on cylindrical rods.

Applying Torque to a Cylindrical Rod

Rotating a cylindrical rod requires the application of torque, which brings a number of factors into play. The first is the polar moment of inertia of a cylindrical rod.

Practical Math Problem

14.6 Polar Moment of Inertia (Hollow Rod)

A hollow rod has an inner radius of 3 centimeters and an outer radius of 5 centimeters. What is its polar moment of inertia?

➤ r = inner radius in meters

➤ R = outer radius in meters

➤ I = polar moment of inertia in meters^4

$$I = \frac{\pi}{2}(R^4 - r^4)$$

The polar moment of inertia is 8.54 x 10^{-6} meters^4.

This formula also applies to solid rods, with r = 0. It is also useful in problems involving shear stress on cylindrical rods (see subsequent problems).

Entering *moment of inertia* into a search engine will reveal a multitude of information on the moments of inertia of various shapes.

One of the most important questions to be answered about cylindrical rods is the maximum amount of torque they can withstand before breaking.

Practical Math Problem

14.7 Maximum Torque for a Cylindrical Rod

For the hollow rod specified in problem 14.6, what is the maximum possible torque that can be applied to the rod if the shear stress cannot exceed 50 megapascals?

> ➤ I = polar moment of inertia (computed as in problem 14.6)
> ➤ R = outer radius in meters
> ➤ S = shear stress limit in pascals
> ➤ T = maximum torque in newton meters

$$T = \frac{SI}{R}$$

The maximum torque is 8,540 newton meters.

When torque is applied to one end of a cylindrical rod and the other end is held fixed, the rod twists. The twist angle can be computed from the rod geometry and the material used in constructing the rod.

Practical Math Problem

14.8 Twist Angle for a Cylindrical Rod

Copper has a shear modulus of elasticity of 936,000,000 pounds per square foot. What is the twist angle in radians of a hollow copper rod 2 feet long with an inner radius of 1 inch and an outer radius of 2 inches when a torque of 5,000 foot-pounds is applied?

➤ r = inner radius in feet

➤ R = outer radius in feet

➤ L = length in feet

➤ S = shear modulus in pounds per square foot

➤ T = applied torque

➤ A = twist angle in radians

First compute the polar moment of inertia (as in Problem 14.6).

$$I = \frac{\pi}{2}(R^4 - r^4)$$

We use this to determine the twist angle.

$$A = \frac{TL}{SI}$$

The twist angle is 0.015 radians = 0.86°.

Compressing and Stretching Bars

The results in this section apply not only to cylindrical rods, but to any type of rod whose cross-sectional area remains constant.

Practical Math Problem

14.9 Lateral Strain of a Compressed Bar

A steel bar with a rectangular cross-sectional area of 40 square inches is subject to an axial compression of 200,000 pounds. Steel has a modulus of elasticity of 2.8×10^7 pounds per square inch and a Poisson ratio of -0.29. What is its lateral strain?

➤ A = cross-sectional area in square inches

➤ F = compressive force in pounds

➤ E = modulus of elasticity in pounds per square inch

➤ P = Poisson ratio

➤ S = lateral strain in inches

$$S = \frac{PF}{EA}$$

The lateral strain is 5.18×10^{-5} inches.

For most common materials, the Poisson ratio is negative, as a compressive axial force (commonly viewed as negative) results in an increase of lateral dimensions (viewed as positive). The same formula applies to exerting axial tension (commonly viewed as positive), which results in a decrease of lateral dimensions. There are a few substances (such as Gore-Tex) that have positive Poisson ratios.

A table of Poisson ratios for common substances can be found by typing *Poisson ratio* into a search engine; similarly for the modulus of elasticity.

A similar formula enables us to find the change in length of a bar of uniform cross-section that is subjected to a force.

Practical Math Problem

14.10 Change in Length of a Bar

The modulus of elasticity (a.k.a. Young's modulus) of aluminum is 10,000,000 pounds per square inch. What is the change in length of an aluminum bar 40 inches long with a cross-sectional area of 25 square inches when subjected to a force of 6,000 pounds?

➤ E = modulus of elasticity in pounds per square inch

➤ L = length in inches

➤ A = cross-sectional area in square inches

➤ F = force in pounds

➤ C = change in length

$$C = \frac{FL}{EA}$$

The length of the bar increases by 0.00096 inches.

Workshop Geometry Associated with Cylindrical Rods

This section contains a few common workshop situations that require geometrical solutions when a cylindrical rod rests on or above a slot. A much greater number of such problems can be found in *Machinery's Handbook*.

Practical Math Problem

14.11 Cylindrical Rod Resting in a Slot

A rod of radius 8 inches rests touching two sides of a slot. The width of the slot at the bottom is 3 inches, and the two sides make angles of 34° and 52°. How far above the floor of the slot is the lowest point on the rod?

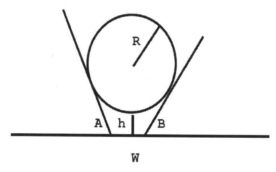

➤ W = width of slot

➤ A, B = angles slot sides make with slot bottom

➤ R = radius of rod

➤ h = height of lowest point of rod above floor

$$h = \frac{R\left(\tan\left(\frac{A}{2}\right) + \tan\left(\frac{B}{2}\right)\right) - W}{\cot A + \cot B}$$

The height of the slot is 1.48 inches.

Alternative Forms

$$W = R\left(\tan\left(\frac{A}{2}\right) + \tan\left(\frac{B}{2}\right)\right) - h(\cot A + \cot B)$$

This formula answers questions such as "What is the width of a slot that makes angles of 32° and 54° with the base of the slot and supports a cylindrical rod of radius 8 inches 2 inches above the base of the slot?" The answer to this question is 1.72 inches.

$$R = \frac{h(\cot A + \cot B) + W}{\tan\left(\frac{A}{2}\right) + \tan\left(\frac{B}{2}\right)}$$

This formula answers questions such as "What is the radius of a cylindrical rod that touches two sides of a 1-inch-wide slot, whose lowest point is inches above the base of the slot, and whose base makes angles of 32° and 54° with the sides of the slot?" The answer to this question is 10.02 inches.

These formulas apply to a rod resting on the floor of the slot by simply letting h = 0.

CHAPTER 15

Moving Stuff from Place to Place

In This Chapter

➤ Simple machines

➤ Pushing and pulling

➤ Transporting fluids

Can I Skip This Chapter?

This chapter deals with the practical math of simple machines used to push and pull things, and the practical math involved in the important business of moving the fluids that are so important to our lives.

Let's face it, a lot of what has to be done in terms of mechanical work involves taking something that is in a place where we don't want it and moving it to someplace we do.

Much of the history of human progress—at least on the engineering front—can be expressed in how our ability to perform these tasks has evolved over time. If Neanderthal man wanted a stone moved from one place to another, he either did it himself or found some buddies to help him. Sometime before the beginning of recorded human history, people discovered that domesticated animals could assist in the performance of this task. At about this time, the first simple machines were invented.

Simple Machines

There are six types of simple machines: the lever, the pulley, the wheel and axle, the wedge, the inclined plane, and the screw. These simple machines are still in use today, and they still perform the basic function of using mechanical advantage to increase the magnitude of a force.

Should misfortune strike you when you are in the middle of nowhere and you get a flat tire, you go to the trunk of your car, unscrew the small tire colloquially known as a doughnut, and get out the jack. A car weighs several tons, but with the aid of a jack, which is basically just a simple machine, even a small child can lift a car. Lifting a car still requires a certain amount of work, but instead of having to do it all at once, you can do it in small increments.

Work is commonly measured in units called foot-pounds—1 foot-pound is the amount of work required to lift a 1-pound brick 1 foot straight up. It is also the same amount of work that is required to lift a 10-pound weight 1/10 of a foot straight up, or a 1,000-pound weight 1/1,000 of a foot straight up. The brilliance of the jack is that even though you must still lift the car, instead of having to lift up hundreds of pounds a couple of feet, the jack enables you to perform the same amount of work by lifting a few pounds a long distance. That's why you need to push the jack handle fairly often, rather than just lift the dead weight of the car straight up a foot or so.

The Lever

The lever is a simple and brilliant idea, so powerful that the Greek scientist Archimedes once stated that if he had a lever and a place to stand on, he could move the world.

Practical Math Problem

15.1 Using a Lever

How much force must be applied to a lever to lift a 2,000-pound weight if the lever is 6 feet long and the fulcrum is positioned 6 inches from the end of the lever closest to the weight?

➤ L = length of lever

➤ W = weight to be lifted

➤ d = distance of fulcrum from weight to be lifted

➤ F = force that must be applied to lift weight

$$F = \frac{dW}{L - d}$$

A force of 181.8 pounds is needed to lift the weight.

The mechanical advantage of the lever in this configuration is (L - d) / d. This is the ratio of the force needed to move a weight to the actual weight.

Alternative Forms

$$d = \frac{FL}{W + F}$$

This formula answers questions such as "How close to the weight to be lifted must the fulcrum be placed for a force of 100 pounds applied to a 5-foot lever to lift a weight of 900 pounds?" The answer to this question is ½ a foot—6 inches.

Of course, in the alternative forms sidebar, if you apply the force by pushing the long end of the lever down 9 inches (¾ of a foot), you will have done ¾ x 100 = 75 foot-pounds of work on the long end of the lever. The same amount of work will have been done at the short end of the lever, and since 75 = 1/12 x 900, the 900-pound weight will only have been lifted 1/12 of a foot—1 inch.

The brilliance of the car jack is that it locks in position after each push of the lever, so instead of lifting the car the same 1 inch every time, the second inch raises the car from a height of 1 inch to a height of 2 inches, etc.

The Inclined Plane

It's easy to push a heavy weight along a relatively frictionless floor; almost no effort is needed. Tilt the floor (or a board under the weight) a little, and very little extra effort is needed. If the board is sufficiently long, you can push a long distance horizontally and move the weight up to an arbitrary height. That tilted board is known as an inclined plane, and that's how the Pyramids and those giant heads on Easter Island were built.

Practical Math Problem

15.2a Using a Frictionless Inclined Plane

How much force is needed to push a 1,000-pound weight up a frictionless inclined plane 10-feet long to a height 2 feet above the ground?

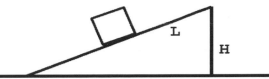

➤ L = length of inclined plane

➤ H = height above floor of end of plane

➤ W = weight of object

➤ F = pushing force needed

$$F = \frac{H}{L}W$$

A force of 200 pounds is needed to move the weight.

The mechanical advantage of an inclined plane is L/H. Again, this is the ratio of the force that must be applied to the weight that is to be moved.

Alternative Forms

$$L = \frac{W}{F}H$$

This formula answers questions such as "How long a frictionless inclined plane is needed to push a weight of 500 pounds up to a height of 3 feet if the pushing force is 100 pounds?" The answer to this question is 15 feet.

The above problem assumes that the inclined plane is frictionless. If the coefficient of friction is denoted by C, then the above formula should be modified as follows.

Practical Math Problem

15.2b Inclined Plane with Friction

$$F = W\left(\frac{H}{L} - C\frac{\sqrt{L^2 - H^2}}{L}\right)$$

In 15.2a and 15.2b, if F is precisely equal to the term on the right-hand side, then it will prevent the block from sliding down the inclined plane but will not push it upward. It needs to be a little more than the quantity on the right-hand side to actually push it up the plane.

The Wheel and Axle

We have come a long way since the days when people got their drinking water from a well, but this was one of the most useful early applications of the wheel and axle.

Practical Math Problem

15.3 Wheel and Axle

A wheel with a radius of 2 feet is concentric with an axle of a radius of 3 inches. A cord is wrapped around the axle and fastened to a 100-pound weight. How much force must be applied to the wheel to lift the weight, and how many revolutions must the wheel turn to lift the weight 3 feet?

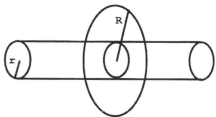

➤ R = radius of wheel

➤ r = radius of axle

➤ W = weight to be lifted

➤ F = force needed to be applied to lift W

➤ D – distance weight must be lifted

➤ N = number of revolutions needed to lift weight

$$F = \frac{rW}{R}$$

$$N = \frac{D}{2\pi R}$$

A force of 12.5 pounds must be applied, and the wheel must be turned 1.91 revolutions.

The mechanical advantage of the wheel and axle is R/r.

In a well, a rope with an attached bucket was wrapped around a wheel, and a handle was cranked to lower the unfilled bucket to below the water level. When the bucket contained water, it would often be a major task to lift it straight up. The mechanical advantage supplied by the wheel and axle enabled a small force to be effectively magnified. The rope wound around the wheel, acting much like a jack in locking in the result of work previously done.

Other Problems Involving Pushing and Pulling

Sometimes simple machines are used in other guises than the straightforward applications we have already looked at, and sometimes we are interested in the amount of effort needed without using simple machines.

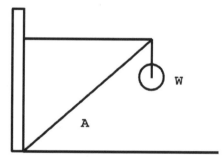

Practical Math Problem

15.4 Forces on a Boom and Cable

A boom of negligible weight makes an angle of 50° with the ground. A cable passes horizontally from a wall to the boom and the cable then suspends vertically to support a weight of 500 pounds. What is the force on the boom and the tension in the cable?

➤ A = angle between boom and ground

➤ W = weight supported by cable

➤ F = force on boom

➤ T = tension in cable

$$F = \frac{W}{\sin A}$$

$$T = \frac{W}{\tan A}$$

The force on the boom is 652.7 pounds. The tension in the cable is 419.5 pounds.

Many fluids, and even some solid material in powdered or gel form, are shipped in cylindrical drums. These sometimes need to be lifted from one height to another. An alternative to direct lifting is to use an inclined plane, or to roll them up a step.

Practical Math Problem

15.5 Rolling a Cylindrical Drum Up a Step

How much force does it take to roll a horizontal cylinder weighing 200 pounds with a radius of 3 feet up a step that is 2 feet high by pushing at the top of the cylinder parallel to the ground?

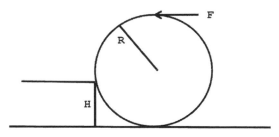

➤ H = height of step in feet

➤ R = radius of cylinder in feet

➤ W = weight of cylinder

➤ F = force in pounds

$$F = \frac{W}{2R - H} \sqrt{2RH - H^2}$$

A force of 141.4 pounds is required.

If H > R it requires less force to simply lift the cylinder straight up than to roll it over the step.

Transporting Fluids

The primary method of transporting fluids from place to place—and one on which civilization depends—is by pipe. Piping moves fluids we want or need, such as water and petroleum, as well as fluids we want to get rid of, such as the waste that travels from our bathrooms to wherever it goes (most of us don't really want to know).

The speed at which fluids move is regulated by pipe geometry.

Practical Math Problem

15.6 Fluid Speed in a Pipe

The cross-sectional area of a pipe at point A is 5 square inches, and the cross-sectional area of the same pipe at point B is 4 square inches. The speed of the fluid flow at point A is 7 feet/second. What is the speed of the fluid flow at point B?

➤ A = cross-sectional area at point A

➤ B = cross-sectional area at point B

➤ a = speed of fluid flow at point A

➤ b = speed of fluid flow at point B

$$b = \frac{aA}{B}$$

The fluid is moving at 8.75 feet/second.

Turbines and Pumps

The turbine is a rotary engine that extracts energy from moving fluids and does useful work with that energy. Turbines power many of our most useful machines, from hydroelectric plants to airplanes.

Practical Math Problem

15.7 Power Output from Fluid Pressure Drop

A flow rate of 50 gallons per minute produces a pressure drop of 20 pounds per square inch across a turbine. What is the power output of the turbine in horsepower?

➤ G = flow rate in gallons per minute

➤ P = pressure drop in pounds per square inch

➤ H = power output in horsepower

$$H = \frac{GP}{1714.3}$$

The turbine produces 0.58 horsepower.

One of the major uses for pumps is to facilitate fluid flow by increasing pressure. Turbines extract energy by decreasing the speed of fluid flow; pumps put energy in and increase the speed of fluid flow.

Practical Math Problem

15.8 Flow Rate of a Pump

A 2-horsepower pump with an efficiency of 80% produces an increase in water pressure of 10 pounds per square inch. What flow rate is necessary to produce this?

➤ E = efficiency in percent

➤ H = pump horsepower

➤ P = pressure differential in pounds per square inch

➤ F = flow rate in cubic feet per second

$$F = \frac{.0382EH}{P}$$

The flow rate is 0.61 cubic feet per second.

Water Pressure Measurement in Head

Water pressure is usually measured in pounds per square inch, but it is also measured in feet of head, which is the number of feet above a specific reference point. The difference in head can be used to measure the pressure difference between two points through which fluid is flowing.

Converting pressure to head can be handled by the following formulas.

Practical Math Problem

15.9 Converting Pressure to Head

Ethyl alcohol has a specific gravity 0.789 and is flowing through a pipe under a pressure of 40 pounds per square inch. What is the head in feet?

> ➤ S = specific gravity of fluid

> ➤ P = pressure in pounds per square inch

> ➤ H = head in feet

$$H = \frac{2.31P}{S}$$

The head is 117 feet.

The following formula handles the conversion in the other direction.

Alternative Forms

$$P = 0.433SH$$

This formula answers questions such as "What is the pressure in a pipe through which water is flowing with a head of 80 feet?" Water has a specific gravity of 1, so the pressure is 34.64 pounds per square inch.

A table of specific gravities of various substances can be found by typing *table specific gravities* into a search engine. Substances with a specific gravity greater than 1 sink in water, substances with a specific gravity less than 1 float on water. In general, a substance with a higher specific gravity than that of a liquid will float in that liquid.

A useful problem is the determination of the speed at which water will flow through a pipe of a given diameter under a given feet of head. This problem is difficult to answer exactly, but the following procedure is generally accurate to within 5% or 10% for level pipes in good condition.

Practical Math Problem

15.10 Velocity of Water Flow in a Pipe

What is the speed at which water under a head of 80 feet is discharged from a pipe that is 4,000 feet long and 2 feet in diameter?

➤ H = head in feet

➤ D = pipe diameter in feet

➤ L = pipe length in feet

➤ V = velocity of discharge in feet per second

Find C from the following table when D is known:

D	C		D	C		D	C
0.1	23		0.8	46		3.5	64
0.2	30		0.9	47		4	66
0.3	34		1	48		5	68
0.4	37		1.5	53		6	70
0.5	39		2	57		7	72
0.6	42		2.5	60		8	74
0.7	44		3	62		10	77

Use the following formula to compute the velocity of discharge.

$$V = C\sqrt{\frac{HD}{L + 54D}}$$

The approximate speed is 11.25 feet per second.

Should you be frequently confronted with this problem, one way to use this formula in a spreadsheet is to make a table with twenty-two rows and five columns. In the example below, these are assumed to be columns A through E and rows one through twenty-two. The first three rows will be

	C	L	H	V
0.1	23			
0.2	30			

The number 30 appears in cell B3, the letter V in cell E1. L and H are the same in all rows, as they are parameters of the problem. Fill in the table values for D in cells A2 through A22 and the corresponding table values for C in cells B2 through B22. In cell E2, type the following formula.

= B2 * sqrt(A2 * D2/(C2 + 54 * A2))

Then highlight cells E2 through E22 and fill down. This will supply the appropriate formula in cells E3 through E22. Check that this gives the correct answer by inserting 4,000 in cell C13 and 80 in cell D13. The number 11.25 should appear in cell E13.

CHAPTER 16

 **Engines and Motors**

In This Chapter

➤ The economics of engines

➤ Work, power, energy, and cost

➤ Engine requirements and design parameters

Can I Skip This Chapter?

The last chapter was about the smaller, simpler machines. This chapter is about the larger machines and is of the most use to people who work with these machines.

Engines and motors are devices for converting energy into mechanical work. Although windmills can convert wind power to electricity and solar mirrors can heat water to make steam to power numerous devices, neither windmills nor solar mirrors have achieved the efficiency and portability to replace the internal combustion engine, which is the workhorse of industry.

There are numerous versions of the internal combustion engine, but the basic idea is simple. Fuel is ignited, and the ignition converts a small volume of fuel to a larger volume of gas at higher temperature and pressure. This moves a component of the engine, which could be a piston or the blades of a turbine, and in so doing produces mechanical work.

That's the way propeller aircraft work. Jet engines and rockets use the expansion of the heated gas directly via Newton's third law, which states that for every action there is an equal and opposite reaction. The heated gas exits the jet or rocket engine in the action portion of Newton's third law, and the jet or rocket moves in the opposite direction as a reaction.

Engine Economics

Once an engine has been built (not necessarily an easy or inexpensive task), there are two basic aspects to its economics. The first is the inherent efficiency of the engine, which is a measure of the fraction of energy input to the engine that is transformed to useful work.

Engine Efficiency

Practical Math Problem

16.1a Engine Power

A gasoline engine burns 3 gallons of gasoline per hour. Gasoline has a heating value of 20,000 BTU/pound and a specific gravity of 0.8. What is the input power of the engine in kilowatts? (1 BTU is one British thermal unit, the amount of heat needed to raise 1 pound of water 1 degree Fahrenheit at 1 atmosphere pressure).

➤ B = fuel heating value in BTU/pound

➤ G = fuel consumption in gallons per hour

➤ S = specific gravity of fuel

➤ P = input power in kilowatts

$$P = \frac{BGS}{410}$$

The input power is 117 kilowatts.

Practical Math Problem

16.1b Engine Efficiency

The engine in problem 16.1a produces an output power of 30 kilowatts. What is its efficiency?

➤ I = input power

➤ O = output power

➤ E = efficiency

$$E = \frac{O}{I}$$

The efficiency of the engine is 0.256. Efficiencies are usually rounded off to two decimal places.

Alternative Forms

$$O = EI$$

This formula answers questions such as "What is the output power of an engine whose efficiency is 0.24 and whose input power is 70 kilowatts?" The answer to this question is 16.8 kilowatts.

Automobile engine efficiencies are generally in the range of 0.25. The first law of thermodynamics limits engine efficiencies to 1.0, as an engine with higher efficiency would produce more power than was fed into it. All claims to have produced such an engine have been found to be in error.

Even though 1.0 is the theoretical limit for engine efficiency, in reality the second law of thermodynamics limits this efficiency even further.

Practical Math Problem

16.2 Limit of Engine Efficiency

What is the maximum possible efficiency of an engine in which heat enters the engine at 1,200°C (centigrade) and exits the engine at 300°C?

➤ T_{in} = temperature at which heat enters engine in degrees centigrade

➤ T_{out} = temperature at which heat exits engine in degrees centigrade

➤ E = maximum engine efficiency

$$E = 1 - \frac{273 + T_{out}}{273 + T_{in}}$$

The maximum efficiency of such an engine is about 0.61.

The awkward appearance of the number 273 in this formula is the result of the conversion of centigrade temperatures to kelvin (absolute temperatures). If temperatures are measured in an absolute scale such as the kelvin or Rankine scales, the formula becomes much simpler and more attractive.

$$E = 1 - \frac{T_{out}}{T_{in}}$$

Formulas for converting between the various temperature scales will be given in Chapter 25.

Fuel Economics

Fuel economics depend upon the energy content of the fuel and the cost of the fuel. Here is a table of the energy content of various fuels, as measured in BTUs:

Coal	BTUs
1 lb	10,000–15,000
1 Ton	25,000,000 (approx.)

Electricity	BTUs
1 kW	3,412

Oil	BTUs
1 Gal #1 fuel	136,000
1 Gal #2 fuel	138,500
1 Gal #3 fuel	141,000
1 Ga. #5 fuel	148,500
1 Gal #6 fuel	152,000

Gas	BTUs
1 lb of butane	21,300
1 Gal of butane	102,600
1 Cu ft of butane	3,260
1 Cu ft of manufactured gas	530
1 Cu ft of mixed gas	850
1 Cu ft of natural gas	1075
1 lb of propane	21,600
1 Gal of propane	91,000
1 Cu ft of propane	2,570

Ethanol does not appear in this table, but ethanol has roughly two-thirds of the energy content of gasoline. This has enormous economic implications, as it means that ethanol needs to be priced at two-thirds the price of gasoline in order to deliver an equal BTU bang for the buck. This does not take into effect factors beyond cost per BTU—possibly ethanol is in the national interest, and possibly it produces different wear in engines capable of using either fuel. However, studies have been done that indicate that if every automobile in the United States were to be fueled by ethanol, most of the land area in the United States would be needed to produce corn to make ethanol.

Another important question is how energy consumption is related to the power of the engine being used.

Practical Math Problem

16.3a Energy Consumption and Engine Power

What is the energy used in kilowatt-hours by a 3-horsepower compressor that runs for 2 hours?

➤ H = engine horsepower

➤ T = running time in hours

➤ E = energy used in kilowatt-hours

$$E = 0.746HT$$

The compressor uses 4.47 kilowatt-hours.

Notice that this formula says that up to a constant factor energy is the product of horsepower multiplied by time. That means that power—and horsepower is a measure of power—is the quotient of energy divided by time. Power is a measure of how fast energy is being used.

This problem also enables us to compute the cost—in money—of running a machine of a given horsepower for a number of hours.

Practical Math Problem

16.3b Cost of Running an Engine

What is the financial cost of running the compressor in problem 16.3a if electricity costs 12¢ per kilowatt-hour?

➤ K = cost per kilowatt-hour

➤ H = engine horsepower

➤ T = running time in hours

➤ C = total cost

$$C = 0.746KHT$$

It costs 54¢ to run the compressor.

To give you some idea of the size of such a compressor, a strong room air conditioner may have a ¼-horsepower motor.

Work, Energy, Power, and Money

As we learned in the last chapter, work is measured in foot-pounds. The power of engines is frequently measured in horsepower. Of course, horses do a miniscule amount of the work in today's industrialized society, but they did a lot in the latter portion of the eighteenth century, when the concept of horsepower was initially developed. Other measurement concepts—inch, foot, yard, pound—trace their lineage even further back in time.

Conversions between all these systems of measurement are extremely important. First of all, work is measured in foot-pounds. Energy can also be measured in foot-pounds, so power, which is energy per second, is also the amount of work done per second. As we shall see in the next chapter, the computation of the amount of energy needed to do a job is frequently easier if we compute the amount of work.

The following formula gives the power requirements for doing a job in terms of the horsepower of the engine that is needed.

Practical Math Problem

16.4 Horsepower Needed to Do Work

What horsepower is delivered by an engine that does 1,000 foot-pounds of work per second?

➤ W = work done in foot-pounds per second

➤ H = horsepower

$$H = \frac{W}{550}$$

The engine must deliver 1.82 horsepower.

Belts are often used to transmit energy. Too strong a force on a belt will cause the belt to break, tear, or snap. This quantity is defined as the allowed pull on a belt. It is measured in terms of pounds per inch of width of the belt.

Practical Math Problem

16.5 Horsepower Transmitted by a Belt

What is the horsepower transmitted by a belt 12 inches wide with an allowed pull of 80 pounds per inch of width and a belt speed of 4,000 feet per minute?

➤ P = allowed pull in pounds per inch of width

➤ W = width in inches

➤ S = belt speed in feet per minute

➤ H = horsepower transmitted by belt

$$H = \frac{PWS}{33,000}$$

The belt transmits 116.4 horsepower.

This is one of those formulas in which almost every solution for every one of the variables is useful. That's because the job that must be done often defines the horsepower needed to do that job, and the problem is to compute one belt parameter in terms of that horsepower and the other belt parameters.

Alternative Forms

$$P = \frac{33,000H}{WS} \qquad W = \frac{33,000H}{PS} \qquad S = \frac{33,000H}{PW}$$

The first formula answers questions such as "What is the allowed pull on a 3-inch-wide belt moving at 500 feet per minute and transmitting 2 horsepower?" The answer to this question is 44 pounds per inch.

This formula answers questions such as "What is the width of a belt with an allowed pull of 50 pounds per inch if it is to move at 800 feet per minute and transmit 1.5 horsepower?" The answer is about 1.24 inches.

This formula answers questions such as "What is the speed with which a 5-inch-wide belt with an allowed pull of 60 pounds per inch must move in order to transmit 0.75 horsepower?" The answer to this question is 82.5 feet per minute.

Requirements for Motors

One of the most important jobs that motors perform is lifting things. The following formula computes the speed at which motors can lift.

Practical Math Problem

16.6 Lifting Velocity of a Motor

A 10-kilowatt engine has an efficiency of 85%. With what constant speed can it lift a 1,000-pound weight?

➤ P = engine power in kilowatts

➤ E = efficiency of engine in percent

➤ W = weight to be lifted in pounds

➤ V = lifting velocity in feet per second

$$V = \frac{7.37PE}{M}$$

The motor lifts at a constant velocity of 6.26 feet per second. This would also serve as the average lifting velocity, assuming that the velocity doesn't vary too much.

Alternative Forms

$$P = \frac{0.136MV}{E}$$

This formula answers questions such as "What power in kilowatts is needed in an engine with an efficiency of 92% in order to lift a 400-pound weight at a constant speed of 5 feet per second?" The answer to this question is 2.96 kilowatts, or about 4.74 horsepower; a kilowatt is approximately 1.36 horsepower.

The remaining formulas in this chapter deal with the design parameters of motors.

Practical Math Problem

16.7 Smallest Rotor Diameter for Motor

What is the smallest diameter for a rotor shaft for a 10-horsepower motor that will rotate at 6,000 rpm and will be subject to a maximum shear stress of 5,000 pounds per square inch?

➤ H = horsepower of motor

➤ R = rotational speed in rpm

➤ S = maximum shear stress in pounds per square inch

➤ D = smallest diameter for rotor shaft in inches

$$D = 68.48 \sqrt[3]{\frac{H}{RS}}$$

The computed answer is 0.475 inches. However, rotor shafts may come in fixed diameters rather than ones that are custom fabricated. If so, a rotor diameter of 0.5 inches is the smallest that can safely be used.

Practical Math Problem

16.8 Wall Thickness to Withstand Pressure

A thin-walled cylinder with a 30-inch diameter is to be made of steel, which has a yield stress of 30,000 pounds per square inch. It must withstand pressures of 500 pounds per square inch and have a safety factor of 3. How thick must the cylinder wall be?

➤ D = diameter in inches

➤ Y = yield stress in pounds per square inch

➤ P = pressure in pounds per square inch

➤ S = safety factor

➤ T = thickness of cylinder wall in inches

$$T = \frac{SPD}{2Y}$$

The cylinder wall must have a thickness of 0.75 inches.

This problem is obviously a significant one for the design of internal combustion engines utilizing pistons enclosed in cylinders.

Entering *yield strengths* into a search engine may return a table with the yield strengths given in megapascals. One megapascal is approximately 145 pounds per square inch.

The next chapter will investigate the work requirements of a number of specific jobs.

CHAPTER 17

Work Needed for Specific Problems

In This Chapter

➤ Heating

➤ Lifting and compressing

➤ Transferring material out of containers

Can I Skip This Chapter?

This chapter continues the theme begun in the previous chapter. The previous chapter concentrated primarily on the relationship between work, energy, power, and cost, although it did discuss something of the design requirements for engines and motors. Anyone who needs to compute the energy requirements of doing jobs requiring mechanical work in its various forms will find this chapter useful.

Work Done Heating, Lifting, and Compressing

The emphasis in this chapter is on the amount of work needed to perform certain specific tasks.

Work and Power in Heating

One of the most important forms of energy is heat, and computing the energy cost of performing certain heating tasks is an important part of daily life.

Practical Math Problem

17.1 Heating Water

It requires 4.18 joules to raise the temperature of 1 gram of water 1°C. How much heat is required to raise 500 grams of water from a temperature of 40°C to 80°C?

➤ t = initial temperature (degrees centigrade)

➤ T = final temperature (degrees centigrade)

➤ c = specific heat (in joules per gram per degree centigrade)

➤ m = mass (in grams)

➤ Q = amount of heat needed (in joules)

$$Q = mc(T - t)$$

It requires 83,600 joules to heat the water. That may sound like a lot, but a joule is a very small amount of heat; 1 kilowatt-hour is 3,600,000 joules. At 12¢ a kilowatt-hour for electricity, it requires a little more than a quarter of a penny to do the job.

Typing *specific heats* into a search engine will locate tables with the specific heats of a large number of materials.

Sometimes it's not enough to simply supply enough heat—it's important to supply it in a short period of time.

Practical Math Problem

17.2 Rate of Temperature Change from Heating

Water has a specific heat of 4.18 joules per gram per degree centigrade. If 500 grams of water are heated with a 3-kilowatt heater, at what rate will the water temperature rise?

➤ c = specific heat in joules per gram per degree centigrade

➤ m = mass of substance in grams

➤ W = power of heater in watts

➤ N = number of degrees centigrade per second

$$N = \frac{W}{cm}$$

The water temperature increases at a rate of 1.44°C per second.

The units for this problem have been chosen because many tables for specific heats of substances give this parameter in joules per gram per degree centigrade. If the mass is given in pounds and the answer is desired in degrees Fahrenheit per second, it is simplest to multiply pounds by 454 to convert to grams, obtain the answer in degrees centigrade per second, and multiply this by 1.8 to obtain the number of degrees Fahrenheit per second.

Lifting a Weight with a Cable

It is easy to compute the work done lifting a weight by itself. Simply multiply the weight in pounds by the number of feet it has been lifted. However, when the cable doing the lifting is not weightless, the problem is a little more complicated.

Practical Math Problem

17.3 Lifting a Weight with a Cable

An 800-pound weight is attached to the bottom of a cable that weighs 4 pounds per foot and goes over the edge of a building. If the weight is initially 200 feet from the top of the building, how much work is done pulling the cable and weight up 10 feet?

➤ W = weight in pounds

➤ L = length of cable (200 feet in above problem)

➤ d = linear density of cable

➤ D = distance that weight is lifted

➤ T = total work done in foot-pounds

$$\mathbf{T \; = \; D(W + d(L - 0.5D))}$$

The total work done is 195,000 foot-pounds.

The computation in this problem assumes that once a section of cable has reached the top of the building, it is no longer lifted but dragged horizontally, which essentially requires no work.

Work Done Compressing a Gas

Practical Math Problem

17.4 Work Done in Reversible Compression

If 100 grams of nitrogen are reversibly compressed from 300 cubic centimeters to 100 cubic centimeters at a constant temperature of 25°C, how much work is done in the process?

➤ R = ideal gas constant = 1.98 calories per gram-mole per degree centigrade

➤ M = molecular weight of gas

➤ m = mass of gas in grams

➤ T = temperature in degrees K

➤ I = initial volume

➤ F = final volume

➤ W = work done in calories

$$W = mT\frac{R}{M}\ln\left(\frac{I}{F}\right)$$

The work done in compressing the gas is 2,354 calories. Calories are the natural units of work (or energy) associated with problems involving heat. Chapter 25 contains the factors needed to convert calories to other units of work.

A reversible compression or expansion is one in which infinitesimal changes can be made to the system without dissipation of energy. Using energy from an explosion is not a reversible process.

Work Done Transferring Material Out of Containers

This is a standard problem. A container is filled, or partially filled, and it is necessary to transfer the material out of the container. Commonly, the material is a liquid and is removed by pumping, but it could also be solid or granular, as long as the material is uniform.

The work done in general depends upon the density of the liquid, the shape of the container, and the depth of material in the container.

We are fortunate to live in a world in which mechanical work is remarkably cheap. One kilowatt-hour is 2,655,224 foot-pounds. That's pretty amazing when you stop to think of it; lighting a 100-watt bulb for ten hours uses the same amount of energy as lifting a 1-ton weight 1,327 feet.

Practical Math Problem

17.5 Partially Filled Rectangular Boxes

A rectangular tank with a length of 8 feet, a width of 6 feet, and a height of 5 feet is filled to a depth of 3 feet with water, which weighs 62.4 pounds per cubic foot. How much work is done pumping out the tank?

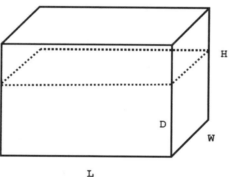

➤ L = length of box

➤ W = width of box

➤ H = height of box

➤ D = depth of material

➤ d = density of material

➤ T = total work done transferring material out of tank

$$T = dLWD(H - 0.5D)$$

The work done was 31,450 foot-pounds.

Practical Math Problem

17.6 Partially Filled Cylinders

A cylindrical tank has a radius of 3 feet and a height of 7 feet. It is filled to a depth of 5 feet with water, which has a density of 62.4 pounds per cubic foot. How much work is done pumping the water out of the tank?

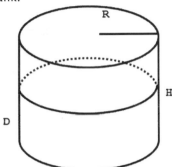

➤ R = radius of cylinder

➤ H = height of cylinder

➤ D = depth of material

➤ d = density of material

➤ W = work done transferring material out of top of tank

$$W = \pi R^2 Dd\left(H - \frac{D}{2}\right)$$

The total work done is 39,697 foot-pounds.

As long as the tank has a constant cross-sectional area A, such as a rectangular tank, a similar formula applies. In this case, the work done pumping out the tank is given by

$$W = ADd\left(h - \frac{D}{2}\right)$$

Practical Math Problem

17.7 Partially Filled Triangular Trough

A tank in the shape of a triangular prism has a length of 10 feet. Its front and back ends are each inverted isosceles triangles with a height of 5 feet and a base of 3 feet. It is filled with water, which has a density of 62.4 pounds per cubic foot, to a depth of 2 feet. How much work is done pumping the water out of the tank?

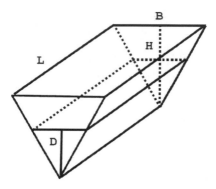

➤ B = base of triangular end

➤ H = height of triangular end

➤ L = length of tank

➤ D = depth of material

➤ d = density of material

➤ W = work done in transferring material out of the tank

$$W = dBLD^2 \frac{3H - 2D}{6H}$$

The work done is 2,746 foot-pounds.

Practical Math Problem

17.8 Partially Filled Inverted Cone

A tank in the shape of an inverted cone with a radius of 4 feet and a height of 10 feet is filled to a depth of 6 feet with water, which has a density of 62.4 pounds per cubic foot. How much work is done pumping the water out of the tank?

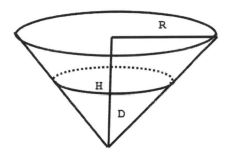

➤ R = radius of cone

➤ H = height of cone

➤ D = depth of material

➤ d = density of material

➤ W = work done transferring material out of top of tank

$$W = \pi R^2 d D^3 \left(\frac{4H - 3D}{12H^2}\right)$$

The work done is 12,421 foot-pounds.

Practical Math Problem

17.9 Partially Filled Hemisphere

A hemispherical tank rests on its base, which has a radius of 5 feet. It is filled to a depth of 3 feet with water, which has a density of 62.4 pounds per cubic foot. How much work is done pumping the water out of the top of the tank?

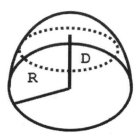

➤ R = radius of hemisphere

➤ D = depth of material

➤ d = density of material

➤ W = work done in transferring material out of the tank

$$W = \frac{\pi d}{4}(R^4 - (R - D)^4)$$

The work done is 29,846 foot-pounds.

PART FOUR

Functional
Math

CHAPTER 18

 **Physics**

In This Chapter

➤ Amount of material: mass, weight, and density

➤ Heat, gravity, sound, and light

➤ Growth and decay

Can I Skip This Chapter?

This chapter contains some of the most fundamental mathematical laws of physics. Most physicists know these formulas, but they may be useful to others, as well as being of interest to people who are curious about what physics is capable of explaining.

The Nature of Physics

Physics is the subject that deals with matter and forces, and the interactions between them. This can be subdivided into a number of different topics such as heat, sound, and light. The laws of physics are one of humanity's most sublime intellectual accomplishments. These laws are based on observation, experimentation, and deduction. Without this knowledge, we would probably still be traveling by oxcart or on horseback, and the only communication we would have over distances would make today's snail mail look lightning fast in comparison.

Mass, Weight, and Density

Mass, weight, and density are basically measures of how much stuff there is. Mass is different from weight. You have the same amount of mass whether you are here on Earth or on the Moon, but you weigh less on the Moon because weight is a measure of how much force gravity exerts on your mass. Density is a relative measure of how much things weigh: iron is denser than water because a specific volume of iron weighs more than an equal volume of water.

Specific gravity is related to density. Density is a measure of how much stuff there is in a particular volume of material: the density of cast iron is about 450 pounds per cubic foot. We saw in problems in Chapter 17 that water has a density of 62.4 pounds per cubic foot. The specific gravity of a material is the ratio of the density of that material to the density of water. The specific gravity for cast iron is approximately 450/62.4 = 7.21. There are no units associated with specific gravity, as the ratio would be the same whether we used both densities of cast iron and water in pounds per cubic feet or grams per cubic centimeter.

Practical Math Problem

18.1 Weight of Material

What is the weight of 12 cubic feet of water?

➤ V = volume of material

➤ d = density of material

➤ W = weight of material

$$W = dV$$

The density of water is 62.4 pounds per cubic foot, so the weight is 748.8 pounds. This formula is only applicable if the density is given in terms of the units in which the volume is measured. If the volume is given in cubic feet, the density must be given in pounds per cubic foot.

The density of water is 62.4 pounds per cubic foot, but it is also 1 gram per cubic centimeter. In fact, that's how the gram was originally defined. This implies that the density of a substance in grams per cubic centimeter is the same as its specific gravity. Typing *specific gravity* into a search engine will find tables with the specific gravity of numerous materials.

There is a table of conversion factors for densities in Chapter 25.

Practical Math Problem

18.2 Fluid Pressure at a Given Depth

What is the pressure at the bottom of a swimming pool that is 9 feet deep?

➤ d = density of liquid in pounds per cubic foot

➤ D = depth of liquid in feet

➤ P = pressure in pounds per square foot

$$P = dD$$

The pressure at the bottom of the pool is 562 pounds per square foot.

In using the metric system, the formula to use is P = gdD, where g = 9.8 meters per second per second, which is the acceleration of gravity. The density in kilograms per cubic meter is d, and D is the depth in meters. Kilograms are a measure of mass, whereas pounds are a measure of weight, which incorporates the acceleration of gravity on the Earth in the number, so g must be included in the metric system answer. As we said earlier, your weight on the Moon is different from what it is on Earth, but your mass remains the same.

Heat

We've seen a few problems on heat in the previous chapter. These problems involved the calculation of how much work was needed to accomplish a specific amount of heating. In this section we look at some of the other aspects of heat.

Practical Math Problem

18.3 Thermal Expansion

An aluminum bar 36 inches long is heated from 70°F to 120°F. How much does it expand? The coefficient of thermal expansion of aluminum is 1.23 × 10⁻⁵ per degree Fahrenheit.

> ➤ L = length of bar
> ➤ t = initial temperature (degrees Fahrenheit)
> ➤ T = final temperature (degrees Fahrenheit)
> ➤ c = coefficient of thermal expansion per degree Fahrenheit
> ➤ E = amount of expansion

$$\mathbf{E = cL(T - t)}$$

The bar will expand 0.022 inches.

Since the coefficient of expansion is given as a dimensionless number, the amount of expansion E will be in the same units as the original length L.

Typing *thermal expansion coefficients* into a search engine will locate tables with the thermal expansion coefficients for many different materials.

If you listen extensively to radio, you will frequently hear advertisements for windows that prevent heat loss. You might wonder how much heat can be lost through a window. The following formula enables you to calculate this. If you do it for all the windows in your house, you may be in for a surprise.

Practical Math Problem

18.4 Heat Loss Through a Surface

The temperature differential on two sides of a wall 10 feet high and 20 feet wide on a cold winter day is 70°F. What is the total amount of heat transferred through the wall across 4 inches of fiberglass insulation in a day? The thermal conductivity of fiberglass is 0.0232 BTUs per foot per degree Fahrenheit per hour.

➤ k = thermal conductivity in BTUs per foot per degree Fahrenheit per hour

➤ L = thickness of insulation in feet

➤ T = temperature difference in degrees Fahrenheit

➤ A = area of wall in square feet

➤ t = time in hours

➤ E = total amount of heat transferred in BTUs

$$E = \frac{kTAt}{L}$$

The total amount of heat that is lost through the wall is 70,157 BTUs. That's why you have to keep heating your house on a cold winter day.

Although the numbers in tables of thermal conductivity are sometimes given in watts per meter per degree kelvin, the conversion factor of 0.5779 will convert watts per meter per degree kelvin to BTUs per foot per degree Fahrenheit per hour is given along with the table and can be found in several online tables.

Practical Math Problem

18.5 Equilibrium Temperature

A 500-gram sheet of copper at 40°C is placed in contact with a 200-gram sheet of aluminum at 20°C. The specific heat of copper is 0.385 joules per gram per degree centigrade, and the specific heat of aluminum is 0.897 in the same units. What is the temperature when they equalize?

> ➤ T_1 = temperature of first substance

> ➤ m_1 = mass of first substance

> ➤ c_1 = specific heat of first substance

> ➤ T_2 = temperature of second substance

> ➤ m_2 = mass of second substance

> ➤ c_2 = specific heat of second substance

> ➤ T = equalization temperature

$$T = \frac{c_1 m_1 T_1 + c_2 m_2 T_2}{c_1 m_1 + c_2 m_2}$$

The equilibrium temperature is 30.35°C.

It doesn't matter in what units the specific heats are given because the ratio of one specific heat to another is the same no matter what system of units is used. Typing *specific heats* into a search engine will locate tables for specific heats of a large number of materials.

Practical Math Problem

18.6 Heat Transferred at Equilibrium

Given the data in problem 18.5, how much heat is transferred from the copper to the iron?

> ➤ Q = amount of heat transferred
> ➤ T = equalization temperature (formula given in previous problem)
> ➤ c = specific heat of cooler substance
> ➤ m = mass of cooler substance
> ➤ t = temperature of cooler substance

$$Q = cm(T - t)$$

The amount of heat transferred is 1,856.8 joules, or about 444 calories (1 joule = 0.239 calories).

Notice the similarity between this formula and the one in problem 18.3. Similarities between formulas are common in physics, because often a process that takes place in one environment is similar to a process that takes place in a different environment. This is reflected in the formulas that describe these processes.

Behavior of Gases

In the last chapter we looked at the work done in reversibly compressing a gas or the work done expanding gas. We now look at what happens to the relationship between the physical parameters of the gas—temperature, pressure, and volume—in different ways of compression.

Practical Math Problem

18.7 Isothermal Compression of Ideal Gas

If 100 cubic feet of an ideal gas at 2 atmospheres pressure is compressed isothermally (keeping the temperature constant) to a volume of 40 cubic feet, what is the pressure of the compressed gas?

➤ p = original pressure of gas

➤ v = original volume of gas

➤ P = final pressure of gas

➤ V = final volume of gas

$$P = \frac{pv}{V}$$

The final pressure is 5 atmospheres.

Practical Math Problem

$$V = \frac{pv}{P}$$

This formula answers questions such as "What is the final volume of an original 100 cubic feet of gas that was compressed at constant temperature from an original pressure of 1 atmosphere to a final pressure of 4 atmospheres?" The answer to this question is 25 cubic feet.

This is also known as Boyle's law, and can be written as

PV = k (pressure x volume = constant) or PV = pv.

Boyle's law describes the relationship between pressure and volume when temperature is kept constant. Charles's law describes the relationship between temperature and volume when pressure is kept constant.

Practical Math Problem

18.8 Isobaric Compression of Ideal Gas

If 200 cubic feet of an ideal gas at 50°C is compressed isobarically (keeping the pressure constant) to a volume of 50 cubic feet, what is the temperature of the compressed gas?

➤ v = original volume of gas

➤ t = original temperature of gas (degrees centigrade)

➤ V = final volume of gas

➤ T = final temperature of gas (degrees centigrade)

$$T = \frac{(t + 273)V}{v} - 273$$

The final temperature is 1,219°C.

Practical Math Problem

$$V = \frac{(T + 273)v}{t + 273}$$

This formula answers questions such as "What is the final volume of an original 100 cubic feet of gas that was heated at constant pressure from 20°C to 40°C?" The answer to this question is 106.8 cubic feet.

This can also be written as

$$\frac{V}{T+273} = k \quad \text{(volume / absolute)temperature = constant) or}$$

$$\frac{v}{t+273} = \frac{V}{T+273}.$$

The Effect of Gravity

Gravity is one of the four basic forces in the universe, the others being electricity (or magnetism—they're different incarnations of the same force); the weak force, which is responsible for radioactivity; and the strong force, which holds the nucleus of an atom together.

There's a lot of material in later chapters on electricity because it plays such a predominant role in our lives. There's a very little bit on radioactivity, which will show up later in this chapter, in this book. And there's nothing on the strong force, because even though it's certainly important that the nucleus of an atom stays intact, there's nothing that practical math has to say on the subject for anyone but a physicist.

Gravity, however, does have an impact on our daily lives. When we throw things, it's a good idea to be able to determine in advance where they land.

Practical Math Problem

18.9 Trajectory of a Thrown Ball

From the top of a hill 100 feet above ground level, a ball is thrown upward at an angle of 37°
to the ground with an initial velocity of 120 feet per second. What is its maximum elevation,
how long is it until the ball strikes ground, and what is the horizontal distance from the point
where it was thrown to where it strikes the ground?

➤ g = acceleration of gravity (32 feet per second per second)

➤ v = initial velocity of ball

➤ h = initial elevation

➤ A = angle of elevation

➤ E = maximum elevation

➤ T = time until ball strikes ground

➤ D = horizontal distance traveled

$$E = h + \frac{v^2 \sin^2 A}{2g}$$

$$T = \frac{v \sin A + \sqrt{v^2 \sin^2 A + 2gh}}{g}$$

Once T has been computed, we can compute D.

$$D = vT \cos A$$

The maximum elevation is 181.5 feet. The ball travels a horizontal distance of 449.2 feet and
lands 5.62 seconds after it was thrown.

Another problem that involves a falling object is determining the depth of a well (or an elevator shaft). You drop something and determine the time until you hear the sound. This enables you to compute the depth of the well.

Practical Math Problem

18.10 Determining the Depth of a Well

A rock is dropped into a well and the splash is heard 2 seconds later. If the speed of sound is 1,110 feet per second, how deep is the well?

> ➤ v = speed of sound in feet per second
> ➤ T = number of seconds until the splash is heard
> ➤ D = depth of well

$$D = \frac{1}{64}(-v + \sqrt{v^2 + 64vT})^2$$

The depth of the well is 60.56 feet.

Light, Sound, and Earthquakes

You might wonder why light, sound, and earthquakes have all been grouped under the same heading. They are all examples of wave phenomena, and they are all characterized by a number called intensity. The intensity of a light is how bright it is, the intensity of a sound is how loud it is, and the intensity of an earthquake is how strong it is.

Practical Math Problem

18.11 Relative Intensity of Sound

A shot from a .357 Magnum pistol registers about 165 decibels. How much more intense is this than the sound of a rock-and-roll singer, who registers 140 decibels?

➤ D = decibel level of louder sound

➤ d = decibel level of quieter sound

➤ I = intensity ratio; louder sound to quieter sound

$$E = 10^{M-m}$$

The pistol shot is approximately 316 times more intense than the rock singer.

This formula is similar to the formula for the relative intensity of earthquakes. If you happen to live in a region in which earthquakes occur, the Richter scale for earthquake magnitude is as familiar as the way the severity of an earthquake is described.

Practical Math Problem

18.12 Relative Intensity of Earthquakes

In January 2010, an earthquake in Haiti killed more than 220,000 people and had a magnitude (Richter reading) of 7.0. The March 2011 earthquake in Japan resulted in fewer than 10,000 fatalities and had a magnitude of 8.9. How much more energy was released in the earthquake in Japan than in the Haiti earthquake?

➤ M = magnitude of more severe earthquake

➤ m = magnitude of less severe earthquake

➤ E = energy ratio, more severe earthquake to less severe earthquake

$$I = 10^{(D-d)/10}$$

The Japan earthquake released about 79.4 times as much energy.

California, especially Southern California, is an area with a lot of earthquakes. It also features another type of vibration, made by the people who come to Southern California seeking a career in music. Of course, Southern California is not the only place where this happens. All stringed instruments, a staple of a wide variety of music, rely on the tension in a stretched string to produce a particular note. The following formula is essential for the creation of music using a stringed instrument.

Practical Math Problem

18.13 Frequency of a Vibrating String

A piano wire is 80 centimeters long and weighs 4 grams. At what tension must it be stretched in order to vibrate at middle C (262 hertz)?

➤ L = length of wire in meters

➤ m = mass of wire in kilograms

➤ f = desired vibration frequency in hertz (cycles per second)

➤ T = tension of string in newtons

$$T = 4Lmf^2 \text{ (newtons)} = 0.8992Lmf^2 \text{ (pounds)}$$

The tension in the string should be 879 newtons, or 198 pounds.

Practical Math Problem

$$f = 0.5\sqrt{\frac{T}{Lm}}$$

This formula answers questions such as "What is the fundamental frequency of a string the length of 50 centimeters that weighs 3 grams and is stretched with a tension of 700 newtons?" The answer to this question is 342 hertz.

Growth and Decay

The PERT formula, which was discussed in conjunction with continuous compounding in Chapter 5, is applicable to general problems of growth and decay.

Practical Math Problem

18.14 Radioactive Decay

Technetium-99m (used in some medical tests) has a half-life of 6 hours. The half-life is the time needed for half of a substance to decay to a nonradioactive state. If 80 milligrams of technetium-99m are administered, how much will be left 24 hours later?

➤ N = original amount of material

➤ H = half-life

➤ T = time (note: T and H must be measured in the same units: hours, days, years, etc.)

➤ A = amount of substance remaining (note: N and A must be measured in the same units: grams, ounces, etc.)

$$A = N\, 0.5^{T/H}$$

After 24 hours, 5 milligrams of technetium-99m remain.

This formula can also be used to describe unrestricted growth such as occurs with bacteria in a petri dish or for a population of animals in an environment with unlimited food and no predators. In this instance, the relevant concept is the doubling period (the time needed for the population to double). If D is the doubling period, the formula is

$$A = N\, 2^{T/D}.$$

Practical Math Problem

$$T = \frac{H \ln(\frac{N}{A})}{\ln 2} = 1.4427\, H \ln\left(\frac{N}{A}\right)$$

This formula answers questions such as "How much time is needed before an original amount of 100 grams of a substance with a half-life of 10 years decays to a level of 5 grams?" The answer to this question is 43.22 years.

CHAPTER 19

 # Electricity

---

In This Chapter

➤ Resistance, capacitance, and inductance

➤ Impedance and reactance

➤ Motors and transformers

Can I Skip This Chapter?

This chapter is primarily for people whose living depends upon the knowledge of how electrical circuits perform or who run electrical equipment and need to know the power consumption parameters and costs involved in using them.

The Force That Really Is With You

Virtually all modern conveniences—*modern* meaning practically anything that has become a part of our daily lives since the beginning of the twentieth century—depend on electricity. Lighting, communication, refrigeration, transportation, you name it. Not only is electricity an end in itself, in the way that it powers so much of our lives, it runs the machines that enable the discoveries of the future to enrich our lives.

You don't really gain an appreciation for the critical importance of electricity until you visit an area of the world that doesn't have reliable electrical power or electrical power is taken away from you via a power outage. Even an outage of a few hours makes you feel that you've been plunged centuries back in time, to a primitive world totally unlike the world we are fortunate to live in.

Electrical Parameters

Electricity is delivered in two forms: direct current and alternating current. Direct current is a one-directional flow of electricity and is the way current is delivered from batteries and solar cells. Alternating current features a periodic reversal of the direction of flow of electric charge; it is the way that electricity is delivered from power plants to homes and businesses.

Simple Electrical Circuits

Practical Math Problem

19.1 Ohm's Law

What is the potential difference in volts resulting from passing a current of 20 amperes through a resistance of 5 ohms?

➤ I = current in amperes

➤ R = resistance in ohms

➤ E = potential difference in volts

$$E = IR$$

The potential difference is 100 volts.

Ohm's law is the fundamental relation between current, resistance, and potential. Current is the rate of flow of electrical charge (which is measured in coulombs) and is analogous to the speed of water flowing in a river. Resistance is the opposition of a material to the flow of current. Potential is the difference in charge between two points in a circuit. Higher potential represents a greater ability to drive electric charge, just as a weight dropped from a higher elevation has a greater ability to generate energy when it hits the ground than the same weight dropped from a lower elevation.

Alternative Forms

$$I = \frac{E}{R}$$

This formula answers questions such as "What is the current resulting from a potential difference of 12 volts across a resistance of 4 ohms?" The answer to this question is 3 amperes, commonly abbreviated as amps.

$$R = \frac{E}{I}$$

This formula answers questions such as "What is the resistance if a potential difference of 60 volts generates a current of 10 amperes?" The answer to this question is 6 ohms.

The two most common types of electrical circuits are the series circuit and the parallel circuit. Components connected in series follow one another along a single path, so that current flows from one component to another.

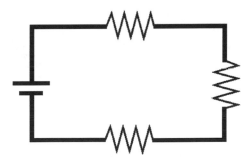

Diagram 19.1 Series Circuit

In a parallel circuit, the current splits its flow, much as a road that offers a choice of which way to go at a junction.

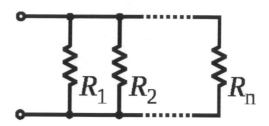

Diagram 19.2 Parallel Circuit

Inductance, Conductance, and Capacitance

The three parameters associated with electrical circuits are inductance, conductance, and capacitance. Conductance is the simplest: it is the reciprocal of resistance so that the lower the resistance the higher the conductance, and vice versa.

A capacitor is a device for temporarily storing charge. Capacitance is the measure of stored charge, the unit used to measure capacitance is the farad (the first five letters of the last name of Michael Faraday, one of the scientists most responsible for the early investigation of electricity).

An inductor generates a magnetic field when current is passed through it. When the current is reduced or stopped (as in a direct current), or changes direction (as in an alternating current), the magnetic field disintegrates, and this causes (or induces) a brief current to flow in the opposite direction. This is known as induction. The strength of the magnetic field is measured in henrys, named after Joseph Henry, another pioneer in the field of electricity.

We frequently find circuits with several resistors, capacitors, or inductors in them. Each resistor (or capacitor or inductor) has a separate resistance (or capacitance or inductance), and we want to know the total resistance (or capacitance or inductance) of the complete circuit.

Practical Math Problem

19.2 Total Resistance in a Circuit

Suppose that three resistors with resistances of 2, 3, and 6 ohms are connected in series. What is the total resistance of the circuit? What is the total resistance of the circuit if they are connected in parallel?

➤ $R_1, R_2, \ldots, R_N$ = resistance in ohms of each resistor

➤ R = resistance of circuit

$$R = R_1 + R_2 + \cdots + R_N \text{ (series)}$$

The resistance of the circuit when all the resistors are connected in series is 11 ohms.

$$R = \frac{1}{\dfrac{1}{R_1} + \dfrac{1}{R_2} + \cdots + \dfrac{1}{R_N}} \text{ (parallel)}$$

The resistance of the circuit when all the resistors are connected in parallel is 1 ohm. The formula for resistance in parallel may be more easily remembered in the following form:

$$\frac{1}{R} = \frac{1}{R_1} + \frac{1}{R_2} + \cdots + \frac{1}{R_N}$$

Practical Math Problem

19.3 Total Capacitance in a Circuit

Suppose that three capacitors with resistances of 10, 15, and 30 farads are connected in series. What is the total capacitance of the circuit? What is the total capacitance of the circuit if they are connected in parallel?

➤ $C_1, C_2, \ldots, C_N$ = capacitance in farads of each capacitor

➤ C = capacitance of circuit

$$C = \frac{1}{\dfrac{1}{C_1} + \dfrac{1}{C_2} + \cdots + \dfrac{1}{C_N}} \text{ (series)}$$

The total capacitance of the circuit when connected in series is 5 farads. As with resistors, the following form is more easily remembered.

$$\frac{1}{C} = \frac{1}{C_1} + \frac{1}{C_2} + \cdots + \frac{1}{C_N}$$

The formula for the capacitance of the circuit when connected in parallel is similar to the series formula for resistors.

$$C = C_1 + C_2 + \cdots + C_N \text{ (parallel)}$$

The capacitance of the circuit when connected in parallel is 55 farads.

Practical Math Problem

19.4 Total Non-Coupled Inductance

Suppose that three non-coupled inductors with inductances of 6, 9, and 18 henrys are connected in series. *Non-coupled* means that the inducting field of one does not affect any of the others. What is the total inductance of the circuit? What is the total inductance of the circuit if they are connected in parallel?

➤ $L_1, L_2, \ldots, L_N$ = inductance in henrys of each non-coupled inductor

➤ L = inductance of circuit

$$L = L_1 + L_2 + \cdots + L_N \text{ (series)}$$

The inductance of the circuit is 33 henrys.

You can probably guess the next formula by analogy with resistance.

$$L = \frac{1}{\dfrac{1}{L_1} + \dfrac{1}{L_2} + \cdots + \dfrac{1}{L_N}} \text{ (parallel)}$$

The inductance of the circuit when all the non-coupled inductors are connected in parallel is 3 henrys. It should not surprise you to know that the following form of this formula for non-coupled inductors in parallel is easier to remember.

$$\frac{1}{L} = \frac{1}{L_1} + \frac{1}{L} + \cdots + \frac{1}{L_N}$$

The situation when the inductors are coupled is more complicated and lies beyond the scope of this book.

Power and Power Dissipation

When current passes through a resistor, some of it is dissipated in the form of heat. This is one of the basic principles used in operating electric heaters. However, when the resistor is part of a circuit that is not being used for heating, there is the potential (not a play on words) for damage to the circuit.

Practical Math Problem

19.5 Power Dissipation in a Resistor

A 5-amp current passes through a 4-ohm resistor. How much power is dissipated?

➤ I = current (in amps)

➤ R = resistance (in ohms)

➤ P = power (in watts)

$$P = I^2R$$

100 watts of power are dissipated.

Alternative Forms

$$P = EI$$

E is the voltage in volts; this is a simple consequence of the fact that E = IR (Ohm's law).

It's important to know how much horsepower an electric motor delivers, just as it is important to know how much horsepower a fueled engine delivers.

Practical Math Problem

19.6 Horsepower of an Electric Motor

What is the horsepower of a 115-volt motor that draws 3 amps of current and operates at an efficiency of 85%?

➤ V = voltage in volts

➤ I = current in amps

➤ E = efficiency (E= 0.85 for a motor with 85% efficiency)

➤ H = horsepower

$$H = \frac{VIE}{746}$$

The motor delivers 0.39 horsepower.

Alternative Forms

$$E = \frac{746H}{IV}$$

This formula answers questions such as "What is the efficiency of a 5-horsepower motor that draws 40 amps at 115 volts?" The efficiency is 0.81, or 81%.

The torque that a motor produces is also important, as we observed in Chapter 14. The term used here is full-load torque, which is the torque a motor produces at its rated horsepower and full-load speed. The next problem explores the relation between the torque and the horsepower.

Practical Math Problem

19.7 Horsepower at Full-Load Torque

What is the horsepower of a motor that runs at 1,800 rpm with a full-load torque of 4 foot-pounds?

➤ R = motor speed (revolutions per minute)

➤ T = full-load torque (foot-pounds)

➤ H = horsepower

$$H = \frac{RT}{5252}$$

The motor delivers 1.37 horsepower.

Alternative Forms

$$T = \frac{5252H}{R}$$

This formula answers questions such as "What is the full-load torque of a 3-horsepower motor that runs at 1,200 rpm?" The answer to this question is 13.13 foot-pounds.

$$R = \frac{5252H}{T}$$

This formula answers questions such as "At what speed will a 2-horsepower motor with a full-load torque of 5 foot-pounds run?" The answer to this question is 2,100 rpm.

Practical Math Problem

19.8 Power Factor (Current and Voltage)

A 21-watt fluorescent lamp ballast draws 21 watts of power off a 120 volt line carrying 0.39 amps. What is the power factor?

Just as efficiency for motors is the ratio of power output to power input, there is a similar definition for electric devices. Here the ratio is called the power factor.

➤ E = voltage (in volts)

➤ I = current (in amps)

➤ P – power (in watts)

➤ F = power factor

$$F = \frac{P}{EI}$$

The power factor is 0.449, which can be regarded as the ratio of real power (what you get out of the device) to the apparent power (which is what comes off the line).

Alternating Current

The parameters for capacitors and inductors in a circuit driven by alternating current differ from those in a circuit driven by direct current. The corresponding parameters are known as capacitative reactance and inductive reactance.

Capacitative reactance, which is a type of resistance to alternating current, is measured in ohms, like resistance. Inductive reactance, like inductance, is measured in henrys.

Practical Math Problem

19.9 Capacitative Reactance

What is the capacitative reactance of a 5,000   F (microfarads) capacitor operating on current with a frequency of 200 hertz?

> ➤ C = capacitance in farads

> ➤ F = frequency in hertz (Hz)

> ➤ X_C = capacitative reactance in ohms

$$X_C = \frac{1}{2\pi FC}$$

The capacitative reactance is 0.16 ohms.

The formula for inductive reactance is simpler.

Practical Math Problem

19.10 Inductive Reactance

What is the inductive reactance of a coil with an inductance of 100   H (microhenrys) operating on current with a frequency of 60 hertz?

> ➤ L = inductance in henrys

> ➤ F = frequency in hertz (Hz)

> ➤ X_L = inductive reactance in ohms

$$X_L = 2\pi FL$$

The inductive reactance is 0.038 ohms. The reactances in problems 19.9 and 19.10 are small because the capacitance and inductance in these problems are on the order of a thousandth of the basic units (farads and henrys).

Impedance

Impedance is another type of resistance to the passage of an alternating current. However, the resistance is not simply a relation between current and voltage, as is the case with Ohm's law in a direct current circuit. Impedance also incorporates the idea of being out of phase in the sense that the crests of one set of waves are to some extent cancelled by the troughs of other sets of waves. In the case of direct current there is no such idea, and so impedance and resistance are one and the same.

In order to calculate impedance, it is necessary to calculate both capacitive and inductive reactances as in problems 19.9 and 19.10. Impedance is a function of resistance and the two reactances, and is also measured in ohms.

Practical Math Problem

19.11 Impedance

What is the total impedance of a circuit that has a 3-ohm resistor, an inductor with an inductive reactance of 5 ohms, and a capacitor with a capacitative reactance of 2 ohms?

➤ R = resistance in ohms

➤ X_L = inductive reactance in ohms

➤ X_C = capacitative reactance in ohms

➤ Z = impedance in ohms

$$Z = \sqrt{R^2 + (X_L - X_C)^2}$$

The impedance of the circuit is 4.24 ohms.

Earlier we saw that in a direct current circuit, the power factor could be calculated from the current and voltage. For an alternating circuit, the power factor can be calculated simply from the resistance and the impedance.

Practical Math Problem

19.12 Power Factor (Resistance and Impedance)

A solenoid valve has a resistance of 970 ohms with an impedance of 1,560 ohms. What is the power factor?

➤ R = resistance in ohms

➤ Z = impedance in ohms

➤ F = power factor

$$F = \frac{R}{Z}$$

The power factor is 0.622.

Electrical Resonance

Electrical resonance is a phenomenon in which the amplitude of the current is larger at certain frequencies than others. These are known as resonant frequencies. They occur in a series circuit when the impedance is at a minimum, or in a parallel circuit when the impedance is at a maximum. In either case, the resonant frequency of an LC circuit (one consisting of capacitors and inductors) can be calculated by the formula in the next problem.

Practical Math Problem

19.13 Resonant Frequency

What is the resonant frequency of an LC circuit with an inductance of 100 H (microhenrys) and a capacitance of 5,000 F (microfarads)?

➤ C = capacitance in farads

➤ L = inductance in henrys

➤ F = resonant frequency in hertz

$$F = \frac{1}{2\pi\sqrt{LC}}$$

The resonant frequency of the circuit is 225 hertz.

Motors and Transformers

One of the primary uses of alternating current is to power motors. A synchronous motor is one in which the speed of the rotation is proportional to the frequency of the alternating current.

Practical Math Problem

19.14 Synchronous Motors

What is the rotation speed of a 2-pole induction motor connected to a 60 hertz power supply?

➤ N = number of poles of motor

➤ F = frequency of power supply in hertz (cycles per second)

➤ R = synchronous rotation speed (rpm) drive

$$R = \frac{120F}{N}$$

The motor is rotating at 3,600 rpm.

One of the main advantages of electrical current is that it can travel long distances and so it can be used far from the place that it was created. However, it travels through resistance, and by problem 9.5 we see that power dissipation is proportional to the square of the current amperage. In order to reduce that dissipation, transformers were invented. A transformer takes current at one amperage and voltage, and changes it into a different amperage and voltage.

When current is generated, the voltage is stepped up by means of a transformer. This reduces the amperage and so reduces the power dissipation in transit. Near the point of use for the current, the reverse transformer steps down the voltage to the 110-volt standard that is in use in the United States.

Practical Math Problem

19.15 Transformer Current and Voltage

A 10-amp current at 110 volts is stepped up to 440 volts by a transformer. What is the output current?

➤ v = incoming (primary) voltage

➤ i = incoming (primary) amperage

➤ V = exiting (secondary) voltage

➤ I = exiting (secondary) amperage

$$I = \frac{iv}{V}$$

The output current is 2.5 amps.

Alternative Forms

$$V = \frac{iv}{I}$$

This formula answers questions such as "What is the output voltage when a 6-amp current coming in at 110 volts is transformed to a 2-amp current output?" The answer to this question is 330 volts.

$$i = \frac{IV}{v}$$

This formula answers questions such as "What input amperage is needed coming in at 110 volts if the output needed is 6 amps at 220 volts?" The answer to this question is 12 amps.

$$v = \frac{IV}{i}$$

This formula answers questions such as "What input voltage is needed if a 6-amp current coming in is to be transformed to an output current of 3 amps at 220 volts?" The answer to this question is 110 volts.

All these formulas are a consequence of the fact that power coming in should be the power coming out: iv = IV. This assumes the transformer operates at an efficiency of 100%. If the transformer efficiency is E (expressed as a percentage), then IV = 0.01Eiv = Eiv/100, and the above formulas should be modified to reflect this. The following formulas are for output.

$$I = \frac{Eiv}{100V}$$

$$V = \frac{Eiv}{100I}$$

The following formulas are for input.

$$i = \frac{100IV}{Ev}$$

$$v = \frac{100IV}{Ei}$$

Transforming current is accomplished by having the input and output current go through coils; the number of turns on the input and output coils determine the outgoing amperage and voltage formulas shown in the next problem.

Practical Math Problem

19.16 Transformer Number of Coil Turns

A transformer has 10 turns in its primary coil and 40 turns in its secondary coil. If the incoming voltage is 110 volts, what will be the outgoing voltage?

➤ v = incoming (primary) voltage

➤ t = number of turns in primary coil

➤ T = number of turns in secondary coil

➤ V = outgoing (secondary) voltage

$$V = \frac{vT}{t}$$

The outgoing voltage is 440 volts.

Practical Math Problem

$$T = \frac{Vt}{v}$$

This formula answers questions such as "How many turns are needed on the secondary coil to step up 110 volts to 220 volts if there are 12 turns on the primary coil?" The answer to this question is 24 turns on the secondary coil.

The above formulas are also valid if amperage is substituted for voltage (i is substituted for v, I is substituted for V).

CHAPTER 20

The Most Economical Way to Do Things

In This Chapter

➤ Maximizing profit and minimizing costs

➤ Taking advantage of geometry

➤ Restocking and processing

Can I Skip This Chapter?

This is a chapter devoted primarily to business-oriented problems such as maximizing profits and minimizing costs from a business standpoint.

Dollars and Sense

The next time you go into a supermarket, take a look at the soup section. Look at the shape of a can of soup. It's a cylinder—but so is a pizza, which has a large radius and a small height, and so is an uncooked spaghetti noodle, which has a small radius and a large height.

There's a very good reason you've never seen a can of soup that's shaped like a pizza or an uncooked spaghetti noodle. Campbell Soup sells on the order of 5 billion cans of soup per year—and what customers buy is the soup, not the can. Campbell Soup needs to use the cheapest possible soup can that contains the necessary amount of soup. Manufacturing soup cans that are shaped like a pizza or an uncooked spaghetti noodle would simply cost too much. Yes, there are other reasons as well: it would be hard to pour the soup out of a pizza-shaped can, and the can shaped like an uncooked spaghetti noodle would be much too long to fit into a grocery bag. However, the big reason is the title of this section. It would cost too many dollars, and that doesn't make any sense. If Campbell's spends a penny too much on each of its 5 billion cans of soup, that's $50,000,000.

The simple version of the soup can problem is the question of what is the shape of the cylinder with a fixed volume (the amount of soup in the can) that has the smallest surface area (the amount of metal needed to make the can). That's a question that practical math can answer.

The real world is more complicated. The material used for sides, top, and bottom is not necessarily the same, and there are costs involved in the choice of manufacturing processes. Nonetheless, the problem of minimizing costs—or maximizing profits—is one that often falls into the realm of practical math, and that's what this chapter is about.

Maximizing Profit and Minimizing Costs

Anyone who runs a business is confronted with the problem of the best price at which to sell their product. The first obvious problem is to determine the total cost of making the product. The only time a business sells below that cost is when it is trying to clear out inventory or when it is introducing a product. In general, though, it is obvious that if you consistently sell below your cost, you'd better find a way to get subsidized or you're going to be out of business.

You can't price the product too high, either. You don't need a course in economics to realize that the demand for a product drops off as the price of that product increases.

Priced too low—not enough profit per item. Priced too high—not enough customers. Often, there's a Goldilocks price: the price that makes the maximum profit.

This is a real-world situation in which it is not possible to come up with a perfect answer in all cases. However, practical math consists of making approximations and estimates that will supply reasonable answers and at least give you a place to start.

In order to determine the Goldilocks price, you need to know something about how demand is related to the price of a product. In many instances, for each extra dollar (or fraction thereof) you charge, you lose roughly the same number of customers. The first problem deals with that situation.

Practical Math Problem

20.1 Maximum Profit from Sales

It costs $30 to manufacture a digital camera. If the sale price is $70, 1 million cameras can be sold. Research shows that for each $1 increase in the sales price, 20,000 fewer cameras will be sold, and for each $1 decrease in the sales price, 20,000 more cameras will be sold. What sales price for the camera will maximize profits?

> ➤ C = cost of manufacture of item

> ➤ P = base sales price ($70 in the above example)

> ➤ N = number of sales at price P

> ➤ n = sales decline for each dollar increase in price

> ➤ S = sales price for maximum profit

$$S = 0.5(\frac{N}{n} + P + C)$$

The digital cameras should be sold for $75.

Of course, in the real world, the numbers N and n are not known exactly, so a company often takes a survey to estimate these numbers, or uses its experience to estimate them.

This problem can also be used to help determine whether a company should actually undertake the manufacture and sale of a product. The results of this problem constitute a best case scenario, and will help a company judge the risk vs. reward of manufacturing the product.

There is a variant of this problem in which a company only has a fixed number of items to sell, and different costs are associated with sold and unsold items.

Practical Math Problem

20.2 Pricing with a Fixed Number of Items

A convention center has space for 300 display booths. It can rent all the display booths if it charges $200 per booth, but for each extra $5 that it charges, one less booth will be rented. It costs $30 to maintain a booth that has been rented, but only $10 to maintain a booth that has not been rented. At what price should it rent its display booths?

➤ N = number of available items

➤ M = maximum price at which all items will be sold

➤ d = dollar increase in price to reduce sales by one unit

➤ s = cost associated with each sold unit

➤ u = cost associated with each unsold unit

➤ P = optimum price at which to sell units

$$P = 0.5(Nd + M + s - u)$$

The convention center should rent display booths for $860 per booth.

Of course, a company that has been doing this for a long time will keep records of how many booths it rents on average at a given price and can compute its maximum profit based on these averages.

For a final look at this particular theme, we turn to a standard problem in agricultural production. If trees (or other plants) are planted too close to each other, there is a competition for available nutrients and sunlight, which reduces the yield per tree.

Practical Math Problem

20.3 Optimum Crop Yield

If 20 apple trees are planted in an acre of ground, they will yield an average of 32 bushels per tree. For each extra tree planted per acre, the average yield per tree is reduced by 2 bushels. How many trees should be planted per acre to obtain the greatest total yield per acre, and what is the optimum yield?

> ➤ N = number of trees planted per acre

> ➤ A = average yield per tree

> ➤ d = decrease in yield for each extra tree

> ➤ T = optimum number of trees to plant per acre

> ➤ Y = optimum yield

$$T = \frac{A + Nd}{2d}$$

$$Y = \frac{(A + Nd)^2}{4d}$$

Plant 18 trees per acre for an optimum yield of 648 bushels per acre.

Packaging in Rectangular Boxes

Unlike Campbell Soup, which packages most of its products in cylinders, many businesses package material in rectangular boxes.

These rectangular boxes can be relatively small for storing and shipping materials. They can also be extremely large, such as rectangular warehouses for storing materials.

Practical Math Problem

20.4 Cheapest Rectangular Box

A rectangular box is to be constructed to contain a volume of 400 cubic inches. Material for the top costs 10¢ per square inch, material for the bottom costs 20¢ per square inch, and material for the sides costs 8¢ per square inch. What are the dimensions of the cheapest such box, and what will it cost?

➤ V = volume of box

➤ t = cost per unit area of top

➤ b = cost per unit area of base

➤ s = cost per unit area of sides

➤ x = length of side of bottom (the cheapest such box will always have a square base)

➤ z = height of box

➤ C = cost of box

$$x = \sqrt[3]{\frac{2sV}{t + b}}$$

Once x has been computed, then z and C can be calculated from the following formulas.

$$z = \frac{V}{x^2}$$

$$C = (t + b)x^2 + 4sxz$$

The base of the box is a square whose side is 5.98 inches and height is 11.2 inches. The cost of the box is $32.13.

If you want to find the box with the smallest surface area and given volume, let $t = b = s = 1$. If you want to find the answer for a box that has no top, let $t = 0$.

Unlike problems 20.1 through 20.3 that assume for each dollar change in price there is a constant drop-off in sales, there are no such assumptions in this problem.

Cylinders and Silos

If you drive through the heartland of America, especially in the grain-producing areas, you will see lots of silos, which are cylinders surmounted by hemispheres. There is obviously a need to construct these as economically as possible.

Practical Math Problem

20.5a Largest Cylindrical Silo

A silo is to be built in the shape of a cylinder surmounted by a hemisphere. The floor of the cylinder costs $10 per square foot, the walls of the cylinder cost $15 per square foot, the surface of the hemisphere costs $25 per foot, and the floor between the cylinder and the hemisphere costs $12 per square foot. A budget of $100,000 is available for construction. What are the dimensions and the volume of the silo with the largest combined volume for cylinder and hemisphere?

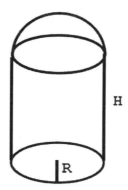

- ➤ f = cost per square foot of cylinder floor
- ➤ F = cost per square foot of hemisphere floor
- ➤ w = cost per square foot of cylinder wall
- ➤ s = cost per square foot of hemisphere surface
- ➤ C = total cost
- ➤ R = radius of hemisphere and cylinder
- ➤ H = height of cylinder
- ➤ V = combined volume of hemisphere and cylinder

$$R = \sqrt{\frac{C}{(3\pi(f+F+2s)-4\pi w)}}$$

Once R has been calculated, R can be calculated from the following formula.

$$H = \left(\frac{f + F + 2s}{w} - 2\right)R$$

Once R and H have been calculated, V can be calculated from the following formula.

$$V = \pi R^2 \left(H + \frac{2R}{3}\right)$$

The radius of the cylinder and hemisphere is 14.28 feet. The height of the cylinder is 40 feet. The combined volume is 31,743 cubic feet.

The formula given in 20.5a can be modified to deal with cylinders without the hemisphere on top by simply letting the costs per square foot associated with the hemisphere equal zero. A similar adjustment will handle the case where there is no hemisphere floor.

Practical Math Problem

20.5b Cheapest Cylindrical Silo

Assume that a silo is to be built with the same parameters as in the previous problem, but that it must have a combined cylinder and hemisphere volume of 20,000 square feet. What are the dimensions of the silo and how much will it cost?

$$R = \sqrt[3]{\frac{3wV}{(3\pi(f + F + 2s) - 4\pi w}}$$

Once R has been calculated, H can be calculated from the following formula.

$$H = \left(\frac{f + F + 2s}{w} - 2\right)R$$

Once R and H have been calculated, C can be calculated from the following formula.

$$C = \pi R^2(f + F + 2s) + 2\pi RHw$$

The radius of the cylinder and hemisphere is 12.25 feet. The height of the cylinder is 34.29 feet. The cost is $73,494.22.

Inventory, Shipping, and Processing

Many companies have central warehouses and a number of stores for selling their products. The products are kept in the central warehouse and shipped to the stores on either a regular schedule or on an as-needed basis. It is important that this be done in a reasonably economical fashion. Too many products of a particular type clog up a store—after all, no drugstore wants to stock only one brand of toothpaste, for example, on its shelves. Too few products mean empty shelves and unsatisfied customers who will go elsewhere for that product, and possibly for other products as well.

Once again, this is a real-world problem that does not have a perfect solution, but practical math can at least come up with a reasonable approximation.

Practical Math Problem

20.6 Smallest Restocking Processing Costs

An electronics store needs to ship 600 big-screen TVs annually from the warehouse to the outlet. It costs $20 to process a single order and $5 to load and ship a single TV. It also costs $15 annually to store a single TV at the outlet. In what size lots should the store reorder and restock TVs in order to minimize total costs?

➤ N = number of items processed annually

➤ P = processing fee for each reorder

➤ C = cost per item for loading and shipping

➤ S = annual storage cost per item

➤ L = optimum lot size

$$L = \sqrt{\frac{2NP}{S}}$$

The optimum lot size to minimize total costs is 40.

In problem 20.6, there is a job of a fixed size to perform: the annual demand for big-screen TVs is known. There is another type of problem in which there is also a job of a fixed size to perform, and the question is how best to utilize labor in the form of persons or machines to do that job. Examples of this type of problem are questions of how many salespeople should be hired to cover a certain territory, or how many machines should be used in order to complete a particular job.

Practical Math Problem

20.7 Optimal Use of Labor

A pretzel-making machine rents for $100 per day and can make 250 pretzels per hour. It costs $50 an hour to hire an operator to supervise the machines. How many machines should be rented to fill an order for 8,000 pretzels and what is the total cost?

➤ N = number of items to be processed

➤ k = number of items that one machine can process in an hour

➤ R = rental costs for one machine per day

➤ H = hourly rate for machine supervision

➤ m = number of machines used

➤ C = total cost

$$m = \sqrt{\frac{hN}{kR}}$$

Once m has been computed, then compute:

$$C = Rm + \frac{hN}{km}$$

The number of machines that should be rented is 4. The total cost will be $800.

Cheapest Way to Lay Pipe or Cable

This is the problem that was referred to in Chapter 1, where the contractor dropped out of the calculus class because he needed the time to fulfill a contract that he had received to wire a building. It isn't the same problem exactly—the real world rarely duplicates exactly what goes on in math books, even practical ones—but practical math gives you guidelines that help give you good answers, even if you can't get perfect ones.

Practical Math Problem

20.8 Cheapest Way to Lay Cable

A bridge crosses a river that is 100 yards wide. Cable must be laid from point A on one side of the river 200 yards from the bridge to the junction B of the bridge with the other side of the river. It costs $800 to lay cable underwater and $500 to lay cable on dry land. The best way is to locate a point C on the other side of the river, lay cable underwater to C, and then along dry land to B. What is the cheapest way to lay the cable, and how much will it cost?

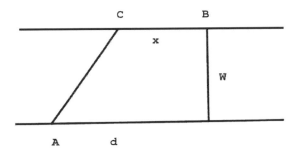

➤ W = width of river

➤ d = distance of point A from bridge

➤ U = cost to lay 1 yard of cable under water

➤ D = cost to lay 1 yard of cable on dry land

➤ x = distance from bridge to point C on other side

➤ C = cost of cheapest cable

There are two cases.

Case 1:

$$x = 0 \qquad\qquad \text{if } U \leq \frac{D\sqrt{W^2+d^2}}{d}$$

$$C = U\sqrt{W^2 + d^2} \qquad\qquad \text{if } U \leq \frac{D\sqrt{W^2+d^2}}{d}$$

Case 2:

$$x = d - \frac{DW}{\sqrt{U^2 - D^2}} \qquad\qquad \text{if } U > \frac{D\sqrt{W^2 + d^2}}{d}$$

Once x has been computed, then compute

$$C = Dx + U\sqrt{W^2 + (d - x)^2}$$

Point C is 120 yards from the bridge, and the cost is approximately \$162,450.

 The Best Way to Do Things

---

In This Chapter

➤ Area and perimeter

➤ Posters, paintings, and windows

➤ Pipes, ladders, and logs

Can I Skip This Chapter?

This chapter continues the quest to find the best way to do things. Most of us have encountered frustrating situations where we misjudged what we needed to purchase. A classic situation is when we purchase a long piece of iron piping that needs to go around an L-shaped corridor—and when we get home, we find that we can't get around the corridor. This chapter helps to resolve some of those situations.

Area and Perimeter

The last chapter was about the cheapest way to do things or how to extract the maximum profit. This chapter deals with the physical world and gets a big assist from geometry.

In the last chapter we wanted to find the cheapest rectangular box that would contain a given volume. We also mentioned that by assuming the top, sides, and bottom all cost 1 unit a piece—it doesn't matter what the units are—you can find the box with the minimum surface area for the given volume. That's a good example of the type of problem that constitutes the bulk of this chapter.

Practical Math Problem

21.1 Largest Rectangle with Fixed Perimeter

What is the largest rectangular field that can be enclosed using 100 feet of fencing, and what are its dimensions?

➤ F = available feet of fencing

➤ L = length of largest field

➤ W = width of largest field

➤ A = area of largest field

$$L = W = \frac{F}{4}$$

$$A = \frac{F^2}{16}$$

The largest rectangle is actually a square with a side of length 25 feet and an area of 625 square feet.

This problem can also be looked at as one of maximizing profit. If you think of the area of the field as being used to grow crops, the largest possible area will also give you the maximum profit from growing something you can sell.

We can also ask about the smallest amount of fencing needed to enclose a rectangular field with a fixed area.

Practical Math Problem

21.2 Smallest Perimeter of Rectangle with Fixed Area

What is the smallest amount of fence that is needed to enclose a rectangular field whose area is 400 square feet, and what are the dimensions of the field?

➤ A = area of field

➤ F = smallest amount of fence

➤ L = length of field

➤ W = width of field

$$\mathbf{L = W = \sqrt{A}}$$

$$\mathbf{F = 4\sqrt{A}}$$

Once again, the rectangle with the smallest perimeter turns out to be a square with a side length of 10 feet. The perimeter requires 40 feet of fence.

This problem can also be looked at as one of doing something in the shortest time. Given that we have to go around a rectangular field of a given area, what are the dimensions of the field that would enable us to do it in the shortest time?

Here are two similar problems involving a field of a different shape: an oval racetrack. Racetracks like this have two semicircles at the opposite ends of a rectangle.

Practical Math Problem

21.3 Greatest Revenue from Oval Display

A convention center is to be constructed in the form of a rectangle with a semicircle at each end. The convention center is to have a perimeter of 1 mile equal to 1,760 yards. The rectangle will produce an average revenue of $50 per square yard, whereas the semicircles will generate an average revenue of $30 per square yard. What should be the dimensions of the convention center to generate the most revenue?

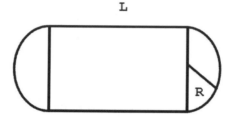

➤ P = perimeter of convention center

➤ r = revenue per square yard for rectangle

➤ s = revenue per square yard for semicircle

➤ L = length of rectangle

➤ R = radius of semicircle

$$R = \frac{rP}{2\pi(2r - s)}$$

Once R has been computed, we can compute L.

$$L = \pi \frac{r - s}{r} R$$

The length of the rectangle is 251.4 yards, and the radius of the semicircle is 200.1 yards.

As we've seen previously, the area can be maximized for a given perimeter by letting r = s = 1.

If this problem parallels 21.1, the next one parallels 21.2

Practical Math Problem

21.4a Minimum Perimeter from Oval Racetrack

A racetrack is to be constructed in the form of a rectangle with a semicircle at each end. If the enclosed area is to be 1,000 square yards, what dimensions of the track will produce the minimum perimeter, and what is that perimeter?

The picture associated with this problem is the one associated with problem 21.3.

> ➤ A = enclosed area
> ➤ L = length of rectangle
> ➤ R = radius of semicircle
> ➤ P = perimeter of track

$$R = \sqrt{\frac{A}{\pi}} \qquad L = 0$$

To minimize the perimeter, the track should be built in the shape of a circle, as circles have the minimum perimeter per unit of enclosed area. Its circumference is 112.1 yards. The radius of the circle should be 17.84 yards.

Practical Math Problem

21.4b Minimum Perimeter from Oval Racetrack

Suppose that in problem 21.4a we require that the area of the rectangle is to be 1,000 square yards. What dimensions of the track will produce the minimum perimeter and what is that perimeter?

$$R = \sqrt{\frac{A}{2\pi}} \qquad L = \sqrt{\frac{\pi A}{2}}$$

Once R and L have been computed, we can compute P.

$$P = 2L + 2\pi R$$

The radius of the semicircle is 12.62 yards, the length of the rectangle is 39.63 yards, and the perimeter is 158.55 yards.

Other Geometrical Considerations

The problems in this section use aspects of geometry other than the area-perimeter relation that was explored in problems 21.1 through 21.4b.

Practical Math Problem

21.5a Smallest Poster with Fixed Print Area

What are the dimensions of the rectangular poster with the smallest area that has side margins of 1 inch, a top margins of 2 inches, a bottom margin of 1.5 inches, and is to contain a rectangular printed area of 60 square inches?

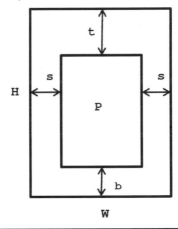

➤ P = printed area

➤ t = top margin

➤ b = bottom margin

➤ s = each side margin

➤ W = width of poster

➤ H = height of poster

$$W = 2s + \sqrt{\frac{2sP}{t+b}}$$

Once W has been computed, we can compute H.

$$H = \frac{t+b}{2s}\,W$$

The width is 7.86 inches and the height is 13.75 inches.

For unequal side margins, substitute the sum of the side margins for 2s in both of the above expressions.

In problem 21.5a, the print area is fixed. It is also possible to solve a related problem in which the overall area and poster margins are fixed.

Practical Math Problem

21.5b Largest Poster Area Inside Margins

A rectangular poster with an area of 300 square inches is to have a top margin of 3 inches, a bottom margin of 2 inches, and side margins of 1 inch each. What are the dimensions of the poster that will yield the largest area inside the margins?

➤ A = area of poster

The rest of the variables are the same as in 21.5a.

$$W = \sqrt{\frac{2sA}{t+b}}$$

Once W has been computed, we can compute H.

$$H = \frac{t+b}{2s} W$$

The width is 10.95 inches and the height is 27.39 inches.

As in 21.5a, for unequal side margins substitute the sum of the side margins for 2s in both of the above expressions.

The following problem may be useful if you ever find yourself wondering where to stand to look at a poster or a painting. The idea is to stand where the painting from top to bottom subtends the largest angle from the standpoint of the observer.

Practical Math Problem

21.6 Where to Stand to View a Painting

A picture 8 feet high is hung so that the bottom of the picture is 2 feet above eye level. At what distance should an observer stand in order to maximize the angle subtended by the picture?

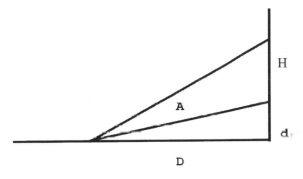

➤ H = height of picture

➤ d = distance of bottom of picture above eye level

➤ A = angle subtended by picture

➤ D = distance from picture of observer

$$D = \sqrt{d(H + d)}$$

The observer should stand 4.47 feet from the picture.

Pipes, Ladders, and Logs

Having seen the type of problem we'll encounter in this book, let's look at another standard problem, and then look at it in some of the other variations in which it appears.

Practical Math Problem

21.7 Carrying a Pipe Around a Corner

What is the longest pipe that can be carried around the corner of an L-shaped corridor that has widths of 6 and 8 feet and is 10 feet high?

➤ D = width of one corridor

➤ d = width of second corridor

➤ H = height of corridor

➤ L = length of longest pipe

$$L = (d^{2/3} + D^{2/3} + D^{2/3})^{3/2}$$

The longest pipe is 41.28 feet.

Admittedly, most of us are not spending time carrying pipes around corners; we carry ladders (which are two-dimensional) or sofas (which are three-dimensional). Even the solution of the two-dimensional problem, such as a ladder, or the footprint of a sofa, is beyond the scope of this book. Nonetheless, computing the answer for a pipe is a good idea; if the diagonal of either a ladder or a sofa is close to the length of the longest possible pipe, there's a very good chance it won't fit. This is especially true of the sofa.

Practical Math Problem

21.8 Shortest Ladder Going Over a Wall

A wall 10 feet high is 16 feet away from a building. What is the shortest ladder that will reach from the ground over the wall to the building?

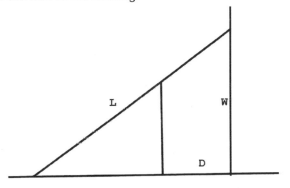

➤ W = height of wall

➤ D = distance from building

➤ L = length of ladder

$$L = (W^{2/3} + D^{2/3})^{3/2}$$

The shortest ladder is 36.44 feet long.

Strongest and Stiffest Beams

Trees generally come with cylindrical trunks. Although cylindrical trunks can be used to construct log cabins, the majority of lumber that is used in the lumber industry is in the form of rectangular beams. The strength of a beam is its ability to resist deflection, which occurs when a load is placed on the beam. The stiffness of a beam is the ratio of the force applied to the deflection produced. A strong beam will support a heavy load; a stiff beam will not bend much when a load is applied.

Practical Math Problem

21.9a Strongest Beam Cut from a Log

What are the dimensions of the strongest rectangular beam that can be cut from a cylindrical log with a diameter of 2 feet?

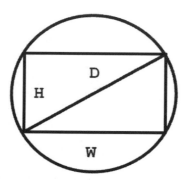

➤ D = diameter of log

➤ H = height of beam

➤ W = width of beam

$$W = \frac{D}{\sqrt{3}} \qquad H = \frac{\sqrt{6}\,D}{3}$$

The width of the strongest beam is 1.15 feet, the height is 1.63 feet.

Practical Math Problem

21.9b Stiffest Beam Cut from a Log

What are the dimensions of the stiffest rectangular beam that can be cut from a cylindrical log with a diameter of 2 feet?

Use the same parameters as in problem 21.9a.

$$W = \frac{D}{2} \qquad H = \frac{\sqrt{3}\,D}{2}$$

The width of the stiffest beam is 1 foot, the height is 1.73 feet.

Rectangles Adjoining a Semicircle

One of the more popular window shapes, especially in Southern California, is the Norman window. This shape, which is also often seen in churches, consists of a rectangle with a semicircle on top. As with all windows, the more light it lets in, the better.

Practical Math Problem

21.10a Most Light Through a Norman Window

A Norman window is a rectangle surmounted by a semicircle. If the glass used for the semicircle transmits 10 lumens per square foot and the glass used for the rectangle transmits 20 lumens per square foot, what are the dimensions of a Norman window with a perimeter of 25 feet that transmits the most light?

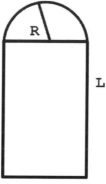

➤ P = perimeter of window

➤ r = intensity of light through rectangle

➤ s = intensity of light through semicircle

➤ L = length of rectangle

➤ R = radius of semicircle

$$R = \frac{r}{2(2 + \pi)r - \pi c}P$$

Once R has been computed, then compute L.

$$L = \frac{P - (2 + \pi)R}{2}$$

The length of the rectangle is 5.21 feet, and the radius of the semicircle is 2.87 feet.

This same calculation can be used if revenue can be generated by the areas within the rectangle and semicircle. See problem 21.3 on maximizing revenue from an oval racetrack.

Practical Math Problem

21.10b Smallest Perimeter of Norman Window

If the total enclosed area of the Norman window of 21.10a is to be 40 square feet, what are the dimensions of the Norman window with the smallest perimeter, and what is the perimeter?

➤ A = enclosed area

$$R = L = \sqrt{\frac{2A}{4+\pi}}$$

Once R and L have been computed, then compute P.

$$P = (4 + \pi)R$$

The radius of the semicircle and length of the rectangle are both 3.35 feet. The perimeter of the window is 23.9 feet. The perimeter is the distance around the Norman window and does not include the side common to the semicircle and rectangle.

CHAPTER 22

Mathematical Tools for Business Management

> ## In This Chapter
>
> ➤ Liquidity and cash flow
> ➤ Asset management
> ➤ Sales evaluation

Can I Skip This Chapter?

This chapter is useful if you own a business, manage a business, or plan on investing in or lending to a business—or have hopes of doing so sometime in the future.

Ratios: Tools for Assessing the Health of Your Business

This chapter is somewhat different from the other chapters in the book. In other chapters, a question is presented, the relevant parameters defined, a formula is given involving those parameters, and an answer is given to the specific question using that formula.

Most of the formulas in this chapter are called ratios. As presented in many texts on business management, they are not ratios, but fractions. Additionally, they are not discussed as if they were fractions, but as percentages.

One such example is the loan-to-value ratio, which has been discussed previously. It is defined as a fraction whose numerator is the amount of the loan and whose denominator is the value. Because most loans that are made are for less than the value, this fraction is less than 1. A typical example might be a loan of $400,000 on a house whose value is $500,000. The actual value of the fraction is 0.8, but this is commonly converted to a percentage by multiplying by 100, and the loan-to-value ratio in this example is described as 80%.

In this chapter, we shall define the more common (and more useful) ratios that occur in the analysis of a business. Some of these ratios are presented in fraction format rather than as a percentage. An example of this is the definition of the loan-to-value ratio as a fraction.

$$\text{loan} - \text{to} - \text{value ratio} = \frac{\$ \text{ amount of loan}}{\$ \text{ amount of value}}$$

To convert the ratio to a percentage, simply multiply by 100.

In the other chapters, a sample question was provided because that's an easy way to determine what math you need in a given situation. These ratios are practical math, but they are not answers to questions in the same way that formulas are; they are assessment tools rather than definitive answers. In order to effectively use the tools presented here, some ratios will be accompanied by a short description of the situations in which they are useful, along with any cautionary information that may be necessary.

This chapter only skims the surface of what is available in the way of ratios that are helpful for business management. A more thorough treatment can be found in the excellent book, *Business Ratios and Formulas: A Comprehensive Guide* by Steven M. Bragg.

Monitor Important Parameters Periodically

Many of the ratios defined in this chapter should be used to monitor the ongoing health of a business by tracking the numbers on a periodic basis (monthly, quarterly, annually) as necessary. If a financial institution tracks its loan-to-value ratio of all loans as 85%, that number on its own may not be especially meaningful. However, if prior computations have resulted in numbers below 80%, the reasons for this change need to be understood. Fractions increase when either the numerators increase or the denominators decrease. In this case, it is possible that an increase in loan-to-value ratio marks an underlying deterioration in the value of the underlying objects on which the loans were made.

You've Got to Break Even Before You Show a Profit

Many—not all—businesses manufacture products. Possibly the single most important thing for a business to know about the product it manufactures is how many it needs to sell in order to break even.

Practical Math Problem

22.1 Business Break-Even Point

The fixed costs for a GPS manufacturer are $3,000,000 per year. It can produce GPS devices for $40, which it sells for $100. Assuming that it sells every GPS it produces, what is its break-even point?

➤ F = fixed costs

➤ C = cost of producing a single unit

➤ S = sales price of a single unit

➤ N = break-even point

$$N = \frac{F}{C - S}$$

The business must sell 50,000 GPS units in order to break even.

The assumption that the company sells every unit it makes for the same price is probably an unrealistic one. There are two other approaches that will probably yield more realistic results.

The first is to estimate the average price at which units will be sold. This approach is more realistic but slightly conservative because it assumes that the sale price for the earlier units will be lower than it probably will be.

A second approach is more accurate mathematically. Make a table that estimates how many units will be sold before the price must be lowered, then how many units will be sold at the lowered price. If you can reasonably estimate these, you will have a more accurate estimate of the break-even point. For the GPS business above, the table might look like this:

Units	Price	Total Net Revenue	Profit
1–30,000	$100	$1,800,000	-$1,200,000
30,001–50,000	$80	$2,600,000	-$400,000
50,001–70,000	$65	$3,100,000	$100,000

The break-even point of 66,000 units can now be found by linear interpolation on the last two rows, as the break-even point clearly lies between 50,001 and 70,000 units. While this method paints a less rosy picture than the simple break-even point computed in problem 22.1, it is less likely to lead to bad decisions.

Liquidity Measurements: Can the Company Survive?

A company's most immediate concern is its ability to pay off its short-term debt. If it cannot do so, it cannot satisfy its creditors. Unless additional funds can be found or credit terms extended, a company failing to pay off short-term debt will likely find itself out of business.

Current, Quick, and Cash Ratios

Measurements of a company's liquidity are not only of interest to the company, they are also important to investors and lenders.

Practical Math Problem

22.2 Current, Quick, and Cash Ratios

➤ A = accounts receivable

➤ S = marketable securities

➤ I = inventory

➤ C = cash

➤ L = current liabilities

$$\textbf{Current Ratio} = \frac{A + S + I + C}{L}$$

$$\textbf{Quick Ratio} = \frac{A + S + C}{L}$$

$$\textbf{Cash Ratio} = \frac{S + C}{L}$$

These are all measures of the company's liquidity. A current ratio of greater than 1 indicates that a company can pay its debts. A cash ratio of less than 1 indicates that it may need to borrow against assets if the liabilities are due in the short term.

The quick ratio is sometimes called the acid test ratio.

Cash Flow Measurements: Has the Company Enough Cash?

While the liquidity measurements presented earlier give a good, quick picture of the health of the company, one of the key questions is whether enough cash is coming in to continue business operations. This section deals with different aspects of this problem.

Cash Flow from Operations Percentage

This number measures the percentage of net income that is generated in cash.

Practical Math Problem

22.3 Cash Flow from Operations Percentage

➤ C = cash flow during period

➤ I = net income during period

$$\text{Cash Flow from Operations Percentage} = 100 \, \frac{C}{I}$$

There are several problems involved in using this easily computed percentage. The first is that there is no uniform definition of cash flow, although the commonly accepted one is cash in minus cash out (Bragg uses net income + noncash expenses – noncash sales). The second is that this percentage will be positive if both the cash flow and the net income are negative. The third is that the percentage itself provides no information as to possible discrepancies between net income and cash flow.

Cash Flow Return on Sales Percentage

This number measures the percentage of total sales that is generated in cash. It can be used, with care, in deciding which product lines are useful to the company in generating cash.

Practical Math Problem

22.4 Cash Flow Return on Sales Percentage

➤ C = cash flow during period

➤ S = total sales during period

$$\textbf{Cash Flow Return on Sales Percentage} = 100 \ \frac{C}{S}$$

This number is only negative when cash flow is negative, remedying one of the deficiencies of the cash flow from operations percentage. However, if a product is near its maximum possible market share, it may not make sense to pour additional resources into expanding that product even though its cash flow return on sales percentage is high relative to other products.

Expense Coverage Days

This is a measure of the number of days that a company can meet its cash obligations with its existing liquid assets. It is especially useful if there is the possibility that income may be temporarily terminated.

Practical Math Problem

22.5 Expense Coverage Days

➤ C = cash

➤ S = short-term marketable securities

➤ A = accounts receivable

➤ E = annual cash expenditures

$$\textbf{Number of Expense Coverage Days} = \frac{C + S + A}{E/365}$$

This is a useful estimate but assumes that all accounts receivable will be paid within the limit specified by the computed value of the expense coverage days. Extending payment periods will result in a decrease in the number of expense coverage days.

Cash Flow to Budgeted Purchases Percentage

This number is used to determine whether a company's cash flow suffices to meet its budgeted obligations. It should be used in conjunction with the computation of expense coverage days.

Practical Math Problem

22.6 Cash Flow to Budgeted Purchases Percentage

➤ C = cash flow during period

➤ F = budgeted fixed asset purchases

$$\text{Cash Flow to Budgeted Puchases Percentage} = 100\frac{C}{F}$$

This number can be skewed by needed one-time purchases. If it is greater than 100, cash must be obtained either through borrowing or the sale of assets.

Performance Measurements: How Is the Company Doing?

The measurements in this section assess the overall health of the company in terms of assets, sales, and profits.

Net Income Percentage

The first number we'll look at is one of the most frequently used measures of company performance. It covers all aspects of the business and sums up how the business is doing.

Practical Math Problem

22.7 Net Income Percentage

➤ I = net income from all sources

➤ R = total revenues

$$\text{Net Income Percentage} = 100\,\frac{I}{R}$$

Companies that make decisions on this index run the risk of having speculative investments made in the hope of inflating the net income percentage. This indeed has proved to be the case in many businesses in the equity and banking industries.

Gross Profit Percentage

This is a key indicator of what the profit percentage for the company actually is, whether it is operating with room to spare if variable costs fluctuate or whether it is walking a narrow line between profit and loss.

Practical Math Problem

22.8 Gross Profit Percentage

➤ R = total revenue

➤ O = total overhead

➤ D = direct material costs

➤ L = direct labor costs

$$\text{Gross Profit Percentage} = 100\,\frac{R - (O + D + L)}{R}$$

Changes in sales volume can affect the gross profit percentage since direct material costs are generally most liable to vary with sales volume while the other costs remain relatively constant as a percentage of revenue. Comparisons can be made between overhead, direct costs, and direct labor to see which factor most heavily impacts the gross profit percentage.

Gross Profit Index

The gross profit index is used to detect fraudulent financial reporting, as this index generally will usually be fairly close to 1 from period to period. While it has value to the company itself, it is especially important to anyone considering investing in the company.

Practical Math Problem

22.9 Gross Profit Index

➤ P_1 = gross profit in period one

➤ S_1 = gross sales in period one

➤ P_2 = gross profit in period two

➤ S_2 = gross sales in period two

$$\textbf{Gross Profit Index} = \frac{P_2/S_2}{P_1/S_1}$$

This index is useful only if the company's fundamental product line remains unchanged from period to period. If the company branches into other product lines, the index may vary greatly from 1 simply because of the different profit margin of the new line.

Gross Margin Percentage

The gross margin percentage is not only a useful indicator in itself, it is also valuable for seeing how much a company marks up its products.

Practical Math Problem

22.10 Gross Margin Percentage

➤ E = total operating expenses

➤ R = total revenue from sales

$$\text{Gross Margin Percentage} = 100\,\frac{R - E}{R}$$

The gross margin percentage is one of the basic indicators of how well a business can survive either a reduction in sales or an increase in operating expenses.

The markup can be deduced from the gross margin percentage (GMP) by means of the following formula:

$$\text{Markup Percentage} = 100\,\frac{GMP}{100 - GMP}$$

Percentage of Operating Assets

This parameter is especially useful to prune out the deadwood in a company's assets. It should be used in conjunction with a detailed breakdown of assets to determine not only how much of the company's assets are nonperforming, but which ones.

Practical Math Problem

22.11 Percentage of Operating Assets

➤ A = current accounts receivable

➤ O_1 = overdue accounts receivable

➤ I = current inventory

➤ O_2 = obsolete inventory

➤ P = production equipment

➤ U = unused equipment

➤ T = total assets

➤ R = revenue-producing assets

$$T = A + I + P + O_1 + O_2 + U$$

$$R = A + I + P$$

$$\text{Percentage of Operating Assets} = 100\frac{R}{T}$$

Overdue accounts receivable may not be operating assets during the period for which this number has been computed, but that does not mean they will stay so indefinitely.

Debt Coverage Ratio

The debt coverage ratio is a more long-term estimate of the company's debt situation than the liquidity measurements presented earlier in this chapter. Tax liabilities are a part of the debt coverage ratio, but they do not enter into the liquidity measurements.

Practical Math Problem

22.12 Debt Coverage Ratio

➤ E = earnings before taxes

➤ T = tax rate as percentage

➤ I = interest expense

➤ P = scheduled principal payments

$$\text{Debt Coverage Ratio} = \frac{E}{I + \dfrac{P}{1 - .01T}}$$

A debt coverage ratio less than 1 (or 100%) indicates that a company may have trouble paying its debts, especially if this continues over more than one accounting period.

How Long to Payback

How long to payback is an important parameter not only for investments by outsiders in a company, but for a company when considering the purchase of capital equipment or other possibly depreciating assets.

Practical Math Problem

22.13 How Long to Payback

➤ I = initial investment

➤ A = average periodic cash flow generated from investment

➤ N = number of periods until payback

$$N = \frac{I}{A}$$

This formula is valid for any period: weeks, months, years. It does not take interest rates into account.

Supporting Sales: Are Sales Commensurate with Assets?

Sales are the means by which a company survives. It is important to know how much of a company's assets are devoted to supporting sales. Too much means that production may suffer; too little may result in a level of sales that's too low.

Sales Expenses to Sales Percentage

This is used to see whether the expenses associated with a particular sales method are excessive. It is sometimes useful to break this number down by product, product line, or sales division.

Practical Math Problem

22.14 Sales Expenses to Sales Percentage

➤ O = overall sales

➤ S = sales salaries plus commissions

➤ E = other sales-related expenses, such as travel

$$\text{Sales Expenses to Sales Percentage} = 100\,\frac{S+E}{O}$$

One concern is that there may be a significant delay between the occurrence of sales and the reporting of sales expenses. This will render the number less useful.

Sales to Working Capital Percentage

This shows the amount of cash necessary to maintain a certain level of sales and can be used to project how much cash will be needed to support changes in levels of sales.

Practical Math Problem

22.15 Sales to Working Capital Percentage

➤ S = annualized net sales

➤ R = accounts receivable

➤ I = inventory

➤ P = accounts payable

$$\text{Sales to Working Capital Percentage} = 100\,\frac{S}{R+I-P}$$

This number may be sensitive to changes in the length of the term for either accounts receivable or accounts payable, which may occur if sales are made to less credit-worthy customers or purchases made from more demanding suppliers.

Sales to Administrative Expenses Percentage

This is used to track the overhead expense needed to maintain a certain level of sales.

Practical Math Problem

22.16 Sales to Administrative Expenses Percentage

➤   S = annualized net sales

➤   E = total general and administrative expenses

$$\text{Sales to Administrative Expenses Percentage} = 100\,\frac{S}{E}$$

This number may change significantly if sales volume changes dramatically, as administrative expenses sometimes change more slowly than sales-related expenses.

Inventory Measurements: Do We Have Enough?

Although it is possible to generate sales without inventory, sooner or later product will have to be delivered to support sales. This section includes parameters that measure the relationship between sales, inventory, and capital.

Sales-to-Inventory Ratio

This ratio is a measure of how much inventory is needed to support sales. This can also be subdivided into inventory categories to see how sales correlate with raw material, goods in process, or finished goods.

Practical Math Problem

22.17 Sales-to-Inventory Ratio

➤ S = annualized sales

➤ I = total inventory (may be subdivided as described above)

$$\text{Sales to Inventory Ratio} = \frac{S}{I}$$

Because both inventory and sales may exhibit seasonal behavior, this ratio is best used by comparing it from one year to another rather than from one quarter to another.

Days of Inventory

Too much inventory indicates that production is outstripping sales; too little indicates that a bottleneck may lie ahead. This measurement should be tracked periodically so the cause for sudden changes may be further analyzed.

Practical Math Problem

22.18 Days of Inventory

➤ C = cost of goods sold

➤ I = inventory

$$\text{Days of Inventory} = \frac{365}{C/I}$$

It is advisable to use average annual inventory in computing the C/I fraction in the denominator to avoid the fluctuations that sometimes result toward the end of an accounting period. The denominator C/I is the number of times inventory is turned over during the course of a year, but using the days of inventory measurement expresses this quantity in number of days of inventory on hand, which is a more intuitive quantity than the number of times inventory is turned over per year.

Inventory-to-Working Capital Ratio

If this ratio is too high, inventory takes up a greater fraction of working capital, which may jeopardize a company's ability to meet short-term cash requirements.

Practical Math Problem

22.19 Inventory-to-Working Capital Ratio

➤ I = inventory

➤ R = accounts receivable

➤ P = accounts payable

$$\text{Inventory to Working Capital Ratio} = \frac{I}{I + R - P}$$

The denominator I + R – P is frequently used as a definition of working capital. Use of this ratio in conjunction with days of inventory as defined previously is advisable, as a company may still be liquid even with a high ratio of inventory to working capital if the inventory turnover is sufficiently brief.

Measurements of Sales Effectiveness: Company and Sales Staff

How well the company's products are selling relative to its competitors, and how well the individual salespeople within the company are doing, are extremely useful management tools.

Market Share Percentage

Tracking market share on a quarterly basis is an effective way of determining how a company is performing within the industry. Increasing market share is the sign of a growing company, declining market share is a cause for concern.

Practical Math Problem

22.20 Market Share Percentage

➤ S = company sales
➤ I = industry sales

$$\text{Market Share Percentage} = 100\,\frac{S}{I}$$

Market share can be artificially inflated during short periods by offering promotions or discounts.

Quote to Close Percentage

The quote to close percentage is an important tool in determining how well a particular salesperson can demonstrate that most vital of sales skills: closing a deal.

Practical Math Problem

22.21 Quote to Close Percentage

➤ R = dollar value of orders received

➤ Q = dollar value of orders quoted

$$\text{Quote to Close Percentage} = 100\frac{R}{Q}$$

This number measures the effectiveness of sales executed by a team, not by an individual.

Pull-Through Percentage

The pull-through percentage can be used in conjunction with the quote to close percentage. The quote to close percentage can be skewed by landing one really big order, whereas the pull-through percentage measures the consistency of closing the deal.

Practical Math Problem

22.22 Pull-Through Percentage

➤ O = number of orders

➤ C = number of contacts

$$\text{Pull} - \text{Through Percentage} = 100\frac{O}{C}$$

Again, when sales are executed by teams, this number can only be used to judge the effectiveness of a team, not an individual. Additionally, it is most valuable when call centers handle the initial contacts, so that there is a record of which individual (or team) first handles the potential customer.

Sales Productivity Percentage

This number can be used to judge the productivity of individual sales personnel or sales teams.

Practical Math Problem

22.23 Sales Productivity Percentage

➤ S = gross revenue from nonrecurring sales

➤ C = sales costs (can vary with different items sold)

➤ E = personnel expenses (salaries, commissions, expenses)

$$\text{Sales Productivity Percentage} = 100\,\frac{S - C}{E}$$

This figure tends to favor low-salary employees because the denominator is smaller. Other measures of gross productivity should be used in conjunction with it.

CHAPTER 23

 **Plane Geometry**

In This Chapter

➤ Quadrilaterals: rectangles, parallelograms, trapezoids

➤ Triangles and other polygons

➤ Circles and ellipses

Can I Skip This Chapter?

Geometry is one of the oldest branches of mathematics and certainly one of the most practical. If it weren't, they wouldn't have been so interested in it so long ago! The only way you won't need to look up some of the formulas in this chapter is if you already know them.

The Importance of Plane Geometry

Almost all the material in this chapter has been around for well more than 2,000 years. In fact, some of it has probably been around for almost 5,000 years, as there is substantial evidence that the earliest civilizations knew more than a little something about plane geometry. They would have had to in order to build the things that they did.

However, formal plane geometry as we learn it in high school traces its origin to the Greeks. They're the ones who worked out the relations among lines, planes, angles, polygons, and circles—and they'd recognize most of the formulas that appear in this chapter.

The formulas in this chapter focus primarily on perimeter and area for familiar shapes.

Quadrilaterals: Rectangles, Parallelograms, and Trapezoids

A quadrilateral is a four-sided polygon. Rectangles are the basic shape in plane geometry. Even though every polygon can be built out of triangles, the rectangle is the fundamental shape for defining and computing area.

Practical Math Problem

23.1 Area and Perimeter of a Rectangle

A rectangle is 12 feet wide and 18 feet long. What are its perimeter and area?

W

L

➤ W = width of rectangle

➤ L = length of rectangle

➤ P = perimeter of rectangle

➤ A = area of rectangle

$$P = 2L + 2W$$

$$A = LW$$

The perimeter of the rectangle is 60 feet. The area is 216 square feet.

Practical Math Problem

23.2 Area of a Parallelogram

What is the area of a parallelogram with a height of 4 feet and a base of 8 feet?

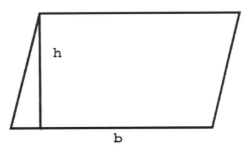

➤ h = height of parallelogram

➤ b = base of parallelogram

➤ A = area of parallelogram

$$A = bh$$

The area of the parallelogram is 32 square feet.

Practical Math Problem

23.3 Area of a Trapezoid

A trapezoid has a lower base of 12 feet, an upper base of 9 feet, and a height of 5 feet. What is its area?

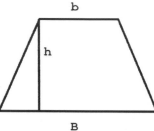

➤ b = length of upper base

➤ B = length of lower base

➤ h = height of trapezoid

➤ A = area of trapezoid

$$A = \frac{(B + b)h}{2}$$

The area of the trapezoid is 52.5 square feet.

Triangles

Although the rectangle is the basic figure for defining area, the triangle is probably the most frequently used shape. Theorems about the lengths of sides and areas of triangles are probably the most useful results in geometry.

Length of a Side of a Triangle

The Pythagorean theorem is probably the most famous theorem in mathematics. Legend says that when Pythagoras proved this theorem, he ordered 100 oxen barbecued to celebrate.

➤ a, b = lengths of sides of right triangle

➤ c = length of hypotenuse

Practical Math Problem

23.4 Hypotenuse of a Right Triangle

What is the length of the hypotenuse of a right triangle with sides of 3 feet and 4 feet?

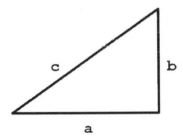

$$c = \sqrt{a^2 + b^2}$$

The hypotenuse of the right triangle is 5 feet long.

➤ a = length of one side

➤ b = length of second side

Practical Math Problem

23.5 Law of Cosines for Triangles

Two sides of a triangle are 6 feet and 8 feet. The angle between the two sides is 34°. What is the length of the third side?

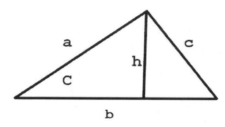

➤ C = size of included angle

➤ c = length of side opposite included angle

$$c = \sqrt{a^2 + b^2 - 2ab \cos C}$$

The length of the third side is 4.52 feet.

Area of a Triangle

There are several different formulas for the area of a triangle, depending upon which lengths and angles are known.

➤ a = length of one side

➤ b = length of second side

➤ C = measure of included an

Practical Math Problem

23.6 Area of a Triangle

23.6a What is the area of a triangle with a base of 10 feet and a height of 12 feet?

23.6b What is the area of a triangle with two sides of 8 and 14 feet and an included angle of 23°?

23.6c What is the area of a triangle with sides that are 3, 4, and 6 feet?

➤ c = length of side opposite included angle

➤ h = height of triangle

➤ s = semi-perimeter of triangle = ½ (a + b + c)

➤ A = area of triangle

$$23.6a \quad A = \frac{bh}{2}$$

$$23.6b \quad A = \frac{ab \sin C}{2}$$

$$23.6c \quad A = \sqrt{s(s-a)(s-b)(s-c)}$$

The area of the triangle in 23.6a is 60 square feet.

The area of the triangle in 23.6b is 21.88 square feet.

The area of the triangle in 23.6c is 5.33 square feet. The formula in 23.6c is known as Heron's Formula.

➤ R = radius of sphere

➤ S = sum of angles (in degrees)

➤ A = area of spherical triangle

Practical Math Problem

23.7 Area of a Spherical Triangle

What is the area of a spherical triangle on a sphere of radius 100 feet with three angles that are 100°, 80°, and 70°?

$$A = \frac{\pi(S - 180)}{180} R^2$$

The area of the triangle is 12,217.3 square feet.

Polygons: Area and Angles

For reasons that are not completely clear, geometrical figures with three straight sides are known as triangles, geometrical figures with four straight sides are known as quadrilaterals, and geometrical figures with any number of straight sides (including three and four) are known as polygons.

➤ A is an interior angle

➤ N = number of sides of polygon

➤ S = sum of interior angles (in degrees)

Practical Math Problem

23.8 Angles in a Polygon

What is the sum of the interior angles in a hexagon (6-sided polygon)?

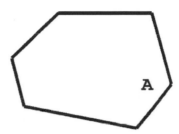

$$S = 180(N - 2) = N - 2 \text{ straight angles}$$

The sum of the interior angles of a hexagon is 720°.

Each interior angle of a regular convex polygon (one with sides of equal length) is 180(N-2)/N degrees.

- ➤ N = number of sides
- ➤ S = length of side
- ➤ A = area of polygon

$$A = \frac{N}{4}S^2\cot(\frac{180}{N}) \quad \text{(angle measured in degrees)}$$

The area is 32.71 square inches.

Practical Math Problem

23.9 Area of a Regular Polygon

What is the area of a regular polygon with 7 sides, each of which is of length 3 inches?

Circles and Ellipses

These two shapes are conic sections (they can be obtained by slicing a cone with a plane), and were well known to the Greeks. The Greeks knew that circles could be used to build wheels, but they were so enamored of circles that they felt that the orbits of the planets were constructed of circles built upon circles. It was later shown by Isaac Newton that the orbits of planets were actually ellipses.

➤ R = radius of circle

➤ C = circumference of circle

➤ A = area of circle

➤ D = diameter of circle

$$C = 2\pi R = \pi D$$

$$A = \pi R^2 = \frac{\pi}{4}D^2$$

Practical Math Problem

23.10 Area and Circumference of a Circle

What are the area and circumference of a circle with a radius of 4 feet?

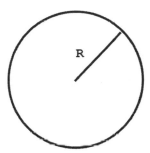

The circumference is 8π (approximately 25.13) feet, and the area is 16π (approximately 50.27) square feet.

> ➤ R = radius of circle
> ➤ θ = central angle in degrees (this is the Greek letter theta)
> ➤ A = area of sector defined by central angle
> ➤ s = arc length subtended by central angle

Practical Math Problem

23.11 Area and Arc Length of a Sector

What is the area of a sector of a circle with a radius of 3 feet with a central angle of 52°, and what is the arc length subtended by that angle?

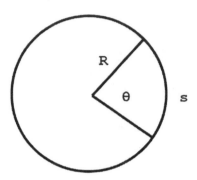

$$A = \frac{\pi R^2 \vartheta}{360}$$

$$s = \frac{\pi R \vartheta}{180}$$

The area of the sector is 4.08 square feet. The length of the subtended arc is 2.72 feet.

➤ R = radius of circle

➤ d = length of chord

➤ θ = central angle in degrees

➤ A = area of circle cut off by chord

➤ s = arc length subtended by central angle

Practical Math Problem

23.12 Area and Angle Cut Off by a Chord

What is the area of the smaller portion of a circle with a radius of 4 feet that is cut off by a chord of 3 feet in length, and what is the central angle subtended by the chord?

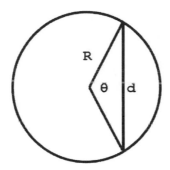

$$A = \frac{\pi R^2}{180} \sin^{-1}\frac{d}{2R} - \frac{d}{2}\sqrt{R^2 - \frac{d^2}{4}}$$

$$\theta = 2\sin^{-1}\frac{d}{2R}$$

The area of the smaller portion of the circle is 0.59 square feet, and the angle subtended by the chord is 44°.

➤ L = length of chord

➤ D = distance from chord midpoint to sector midpoint

➤ R = radius of circle

Practical Math Problem

23.13 Determining the Radius from a Chord

The length of a chord of a circle is 6 inches, and the distance from the center of the chord to the minor arc of the circle it determines is 2 inches. What is the radius of the circle?

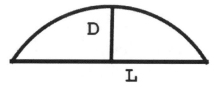

$$R = \frac{d}{2} + \frac{L^2}{8d}$$

The radius of the circle is 3.25 inches.

➤ a = semimajor axis

➤ b = semiminor axis (a > b)

➤ A = area of ellipse

➤ C = circumference of ellipse

Practical Math Problem

23.14 Area and Circumference of an Ellipse

What are the area and circumference of an ellipse with a semimajor axis of 8 inches and a semiminor axis of 6 inches?

$$A = \pi ab$$

This is a complicated formula, and it's simpler if we introduce an abbreviation.

$$u = \frac{a-b}{a+b}$$

The formula below for C is approximate. No simple exact formula exists for the circumference of an ellipse.

$$C \approx \pi(a + b)\left(1 + \frac{3u}{10 + \sqrt{4 - 3u}}\right)$$

The area of the ellipse is 150.8 square inches. Its circumference is approximately 45.6 inches.

CHAPTER 24
Solid Geometry

In This Chapter

➤ Boxes and cylinders

➤ Pyramids and cones

➤ The sphere and the torus (a.k.a. the doughnut)

Can I Skip This Chapter?

You probably won't need the formulas in this chapter as often as you will need the formulas on plane geometry. On the other hand, unless your job requires you to be familiar with the formulas in this chapter, you almost certainly don't know them.

The Importance of Solid Geometry

We live in a three-dimensional world. The Greeks knew that, and while they were busy developing plane geometry, they did not neglect solid geometry. The objects of solid geometry are the objects that comprise the world around us. Things that aren't spheres, or cones, or boxes, or cylinders are often constructed of the basic objects of solid geometry.

The questions in this chapter relate to volume and surface area. There are a lot of practical math questions that need to be answered using volume and surface area.

Problems on how big something must be in order to store a certain amount of material require the formula for volume—think of how ridiculous it would be to build a storage facility that is too small for your needs. It wouldn't be as ridiculous to build a storage facility that's too large, but it would be more expensive than necessary.

Solids with Constant Cross Section

Solids with constant cross section include two of the most important solids that we come in contact with in everyday life: rectangular boxes and cylinders.

Practical Math Problem

24.1 Volume of a Solid with Constant Cross Section

What is the volume of a solid of constant cross section that has a base area of 12 square inches and a height of 16 inches?

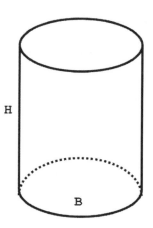 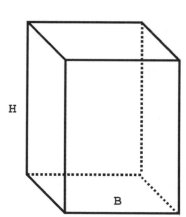

➤ B = area of base

➤ H = height of solid

➤ V = volume of solid

$$V = BH$$

Both of the above solids have a volume of 192 cubic inches. The same would be true no matter what the shape of the base, as long as every cross section parallel to the base has the identical shape.

The next problem is one encountered more often in real life than in geometry books.

Practical Math Problem

24.2 Volume Remaining in a Horizontal Cylindrical Tank

A cylindrical tank that is 10 feet long and has a radius of 4 feet is lying on its side and contains liquid. The liquid level is 1 foot below the top of the tank. What is the volume of the liquid?

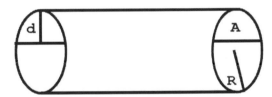

➤ R = radius of tank

➤ L = length of tank

➤ d = distance of liquid level from top of tank

➤ A = cross-sectional area remaining above liquid level

➤ V = volume of remaining liquid

There are two cases: if d < R and if d > R.

Make sure to set the calculator to radian mode!

If d < R, first compute the following value of A.

$$A = R^2 \cos^{-1}\left(\frac{R-d}{R}\right) - (R-d)\sqrt{d(2R-d)}$$

It is now possible to compute the volume.

$$V = (\pi R^2 - A)L$$

The volume of the liquid is 466.4 cubic feet.

If d > R, it will simplify the final formula to first make the following computation.

$$u = 2R - d$$

Now compute the following value for x.

$$x = R^2 \cos^{-1}\left(\frac{R-u}{R}\right) - (R-u)\sqrt{u(2R-u)}$$

We are finally able to compute the volume.

$$V = Lx$$

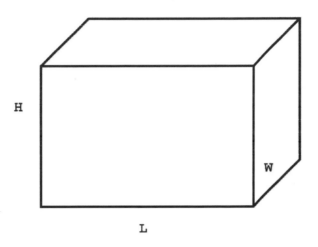

Practical Math Problem

24.3 Surface Area of a Rectangular Box

A rectangular box is 8 inches long, 6 inches wide, and 12 inches long. What is its surface area?

- ➤ L = length of box
- ➤ W = width of box
- ➤ H = height of box
- ➤ A = total surface area of box

$$A = 2WL + 2WH + 2LH$$

The surface area of the box is 432 square feet.

Practical Math Problem

24.4 Surface Area of a Cylinder

What are the lateral surface area and the total surface area of a cylinder that is 10 inches high and has a radius of 4 inches?

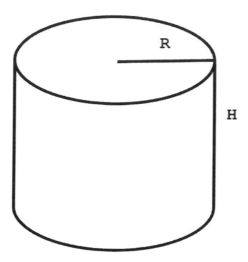

➤ R = radius of cylinder

➤ H = height of cylinder

➤ L = lateral surface area of cylinder

➤ A = total surface area of cylinder

$$L = 2\pi RH$$

$$A = 2\pi R(H + R)$$

The lateral surface area of the cylinder is 251.33 square feet. The total surface area is 351.86 feet.

Regular Pointed Solids

Regular pointed solids include pyramids that have the vertex directly over the center of the base, as well as cones. They have been grouped together because the formulas for both volume and surface are similar.

Practical Math Problem

24.5 Volume of a Regular Pointed Solid

What is the volume of a regular pointed solid with a base that has an area of 50 square inches and a height of 12 inches?

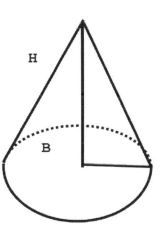 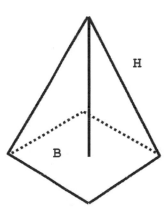

➤ B = area of base

➤ H = height of solid

➤ V = volume of solid

$$V = \frac{1}{3}BH$$

The volume of the solid (whether a cone or a pyramid) is 200 cubic inches.

This formula is similar to the formula for the volume of a solid of constant cross section, as it does not matter what the shape of the base is as long as it satisfies the definition of a regular pointed solid.

Practical Math Problem

24.6 Surface Area of a Pyramid

What are the slant height, lateral surface area, and total surface area of a 4-sided regular pyramid with 3-inch sides and a height of 10 inches?

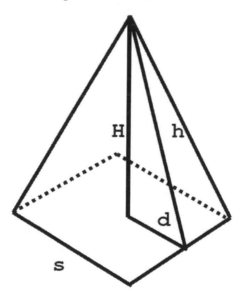

➤ n = number of sides of base (must be a regular polygon)

➤ s = length of side of base

➤ H = height of pyramid

➤ h = slant height of pyramid

➤ d = distance from center of regular polygon to midpoint of side

➤ L = lateral surface area

➤ A = total surface area

$$d = \frac{1}{2}s\cot(\frac{180}{n})$$ angle measured in degrees

Once d has been computed, we can compute the slant height h.

$$h = \sqrt{d^2 + H^2}$$

Once h has been computed, we can compute the surface areas.

$$L = \frac{1}{2}nsh$$

$$A = \frac{1}{2}ns(h + d)$$

The slant height of the pyramid is 10.11 inches. Its lateral surface area is 60.67 inches. Its total surface area is 69.67 square inches.

Practical Math Problem

24.7 Surface Area of a Cone

What are the slant height, lateral surface area, and total surface area of a cone with a radius of 6 inches and a height of 8 inches?

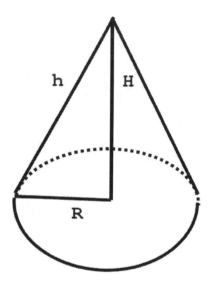

> ➤ R = radius of cone
> ➤ H = height of cone
> ➤ h = slant height of cone
> ➤ L = lateral surface area
> ➤ A = total surface area

$$h = \sqrt{R^2 + H^2}$$

Once h has been computed, we can compute the surface areas.

$$L = \pi Rh$$

$$A = \pi R(h + R)$$

The slant height of the cone is 10 inches. Its lateral surface area is 188.5 inches. Its total surface area is 301.59 square inches.

Spheres

In some ways, the sphere is the simplest solid. Like the cube, one only needs to specify a single parameter (its radius). Spheres are central to our civilization; many machines rely on ball bearings—small spheres of metal—in order to function. Many of our games center upon spheres as well. As Gilbert and Sullivan said in The Mikado, we would not relish playing billiards "on a cloth untrue with a twisted cue and elliptical billiard balls." It's hard to conceive of baseball, tennis, basketball, golf, or soccer using a ball with the fundamental shape other than a sphere.

As storage containers, spheres have the unique property of using the smallest amount of material for a given volume.

Practical Math Problem

24.8 Volume and Surface Area of a Sphere

What are the volume and surface area of a sphere with a radius of 5 feet?

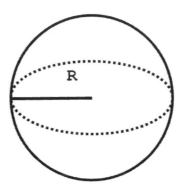

➤ R = radius of sphere

➤ V = volume of sphere

➤ A = surface area of sphere

$$V = \frac{4}{3}\pi R^3$$

$$A = 4\pi R^2$$

The volume of the sphere is 523.6 cubic feet. Its surface area is 314.2 square feet.

Practical Math Problem

24.9 Volume of a Spherical Cap

What is the volume of a spherical cap 3 feet high from a sphere with a radius of 5 feet?

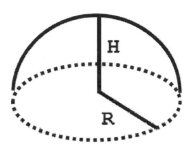

➤ R = radius of sphere

➤ H = height of cap

➤ V = volume of cap

$$V = \pi H^2 \left(R - \frac{H}{3} \right)$$

The volume of the spherical cap is 113.1 cubic feet.

The Torus

If you're in a hurry and need breakfast on the run, a cup of coffee and a doughnut may be just the ticket. Both are fundamental shapes from solid geometry. The coffee will come in a hemispherical ceramic cup, a cylinder-shaped mug, or a paper or plastic cup that is the frustum of a cone (the bottom portion of a cone after the top has been chopped off). It may even come in one of those paper cones that is inserted through a circular plastic cup holder.

The doughnut itself is a torus, a circle revolved around a central axis. Other examples of the torus are tires and inner tubes.

Practical Math Problem

24.10 Volume and Surface Area of a Torus

A torus is formed by rotating a circle with a radius of 3 inches parallel to a central axis located 10 inches from the center of the circle. What are the volume and surface area of the torus?

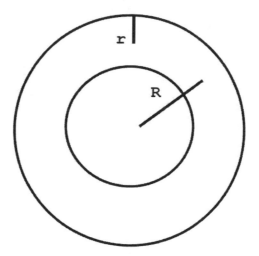

➤ r = radius of rotated circle

➤ R = distance of central axis to center of rotated circle

➤ V = volume of torus

➤ S = surface area of torus

$$V = 2\pi^2 r^2 R$$

$$S = 4\pi^2 rR$$

The volume of the torus is 1,776.5 cubic inches. Its surface area is 1,184.4 square inches.

Although these formulas may look confusing, it's fairly easy to see where they come from. If you imagine that the doughnut is made out of something like rubber, put it on a chopping board and slice it, you could stretch it out and get a cylinder with a base that is the rotated circle of radius r and a height that is the circumference of the circle with a radius of R.

There is a more general version of this result known as the theorems of Pappus that applies to a closed curve that moves in a path parallel to a central axis. In the above example (the torus), the closed curve is a circle. In this case:

➤ P = perimeter of closed curve

➤ L = length of path described by centroid of closed curve

➤ A = area of closed curve

$$V = AL$$

$$S = PL$$

PART FIVE

Math Aids

CHAPTER 25

 **Conversions**

In This Chapter

➤ Simple and complicated conversions

➤ English and metric systems

➤ Tables of conversions

Can I Skip This Chapter?

We live in a world with a variety of different ways of measuring things. Although the information in this chapter is readily available online, it's nice to have the most common ones collected in one place—and this is the place.

Different Systems of Measurement

There are lots of different ways of measuring things, but when two different systems of measurement are used, it is critically important to be able to convert one to the other.

Numbers, by themselves don't really mean anything in relation to the real world. It is the units attached to these numbers that are important. Getting paid 1,000 to do a day's work sounds really good, but only if you are being paid 1,000 in dollars or maybe euros. However, 1,000 Italian lira are only worth about 72¢ at today's rate of exchange.

The rates of currency exchange vary, but the conversions in this chapter do not: 1 kilogram will always be approximately 2.2 pounds, and 1 meter will always be approximately 39.37 inches. This chapter contains almost all the conversion information most people will ever need to know.

Some General Comments about Conversions

The tables in this chapter handle most common conversion problems. However, you will occasionally encounter a problem in which the units are unfamiliar. Here is an example to illustrate different ways of approaching such a problem.

A gallon of liquid energy fuel supplies 20,000 BTUs and costs $5. For how long can one run an appliance that requires 2 kilowatts of energy per hour with a given budget of $100?

Solution 1: This problem would be fairly straightforward if one knew how many BTUs of energy per hour the appliance uses. This suggests that it is desirable to convert kilowatt-hours to BTUs. The Basic Conversions: Heat and Energy table shows that 1 watt-hour is equivalent to 3.412 BTUs, so 2 kilowatt-hours is $2 \times 1,000 \times 3.412 = 6,824$ BTUs. Since $5 purchases 20,000 BTUs, each dollar purchases $20,000/5 = 4,000$ BTUs, and $100 will purchase $100 \times 4,000 = 400,000$ BTUs. Each hour of operation requires 6,824 BTUs, and so the number of hours of operation is $400,000/6,824 = 58.6$ hours.

Solution 2: The Basic Conversions: Heat and Energy shows that 1 BTU is equivalent to 0.293 watt-hours, so 1 gallon of liquid energy supplies 20,000 BTUs = 20,000 x 0.293 watt-hours = 5,860 watt-hours. Since 1 kilowatt-hour is 1,000 watt-hours, 1 gallon of liquid energy fuel supplies 5.86 kilowatt-hours. Since 1 gallon of liquid energy fuel costs $5, the kilowatt-hour production from each dollar is 5.86 kilowatt-hours divided by $5, or 1.172 kilowatt-hours per dollar. Since the budget is $100, the total amount of kilowatt-hours that $100 will purchase is $100 \times 1.172 = 117.2$ kilowatt-hours. The appliance requires 2 kilowatts per hour to operate, so the total time it can operate is $117.2/2 = 58.6$ hours.

A straightforward way to attack a question phrased in the English system, such as, "How many BTUs are required to raise the temperature of 20 pounds of water 10°F?", is to convert the weight and temperature into the metric system, obtain the answer in joules, and convert joules to BTUs using the table in this chapter.

In general, it is easier to handle all problems given in English units when data is supplied in metric units by converting the English unit quantities into metric unit quantities, applying the appropriate formula to obtain the answer in metric units, and then converting back to English units.

Successive Conversions

How many square feet are there in an acre?

There is no conversion factor given in the book for this. However, one can find one conversion factor for acres to square yards (4,840) and one conversion factor for square yards to square feet (9). The conversion factor for acres to square feet is the product

4,840 x 9 = 43,560. One way to remember this is by cancelling the common numerator and denominator in the following equation:

$$\frac{\text{sq. feet}}{\text{sq. yard}} \times \frac{\text{sq. yard}}{\text{acre}} = \frac{\text{sq. feet}}{\text{acre}}$$

(square feet per square yard times square yards per acre = square feet per acre).

Reverse Conversions

How many square miles are there in an acre?

There is no conversion factor given in the book for this. However, one can find one conversion factor for square miles to acres (640). The conversion factor for acres to square miles is the reciprocal 1/640 = 0.0015625. One way to remember this is to recall that division by a fraction is accomplished by multiplying by the inverted fraction; division by a, b is accomplished by multiplying by b, a. So

$$\frac{\text{sq. miles}}{\text{acre}} = 1 \Big/ \frac{\text{acres}}{\text{sq. mile}}$$

(square miles per acre = 1 divided by acres per square mile).

Converting Between the English and Metric Systems

Many conversions between the English and metric systems are a straightforward matter of multiplication by an appropriate constant. This applies to any conversion that is strictly based on the length dimension, such as length, area, and volume: 1 foot is 0.3408 meters, so 3 feet is $3 \times 0.3408 = 1.0224$ meters. Similarly, 1 square foot is 0.0929 square meters, so 3 square feet is $3 \times 0.0929 = 0.2787$ square meters, and 1 cubic inch is 16.387 cubic centimeters, so 3 cubic inches is 49.161 cubic centimeters.

The most critical difference occurs in converting masses and weights. Mass is the amount of stuff a body possesses, whereas weight is the force needed to move that mass. That force depends on gravity, and the gravitational force depends upon the size of the body. You can jump much higher on the Moon than on Earth because even though your mass is the same in both places, your weight on the Moon is only one-sixth of your weight on Earth.

Mass in the metric system is measured in kilograms, and weight in the metric system is measured in newtons. You've never gone into a store and bought several newtons worth of canned tomatoes, so even though the weight of the canned tomatoes is given in both kilograms and pounds, the weight in kilograms is something of a misnomer. This generally doesn't make any difference, except in problems that rely on the difference between mass and weight. One such example is the pressure at a given depth in a fluid. The formula for pressure in the English system just multiplies density times depth, whereas the metric system multiplies density times depth times the acceleration of gravity. Most problems in this book use the English system to avoid possible confusion.

Different Temperature Scales

There are three basic systems for measuring temperature. In the Fahrenheit system, generally used throughout the English-speaking world, ice melts at 32°F and water boils at 212°F. Many countries in which English is not the predominant language use the centigrade (a.k.a. Celsius) system in which ice melts at 0°C and water boils at 100°C. The centigrade system is also commonly used for scientific and engineering measurements.

Finally, the kelvin (a.k.a. absolute) system has as its basic unit the centigrade degree, except that 0°K corresponds to absolute zero; the lowest possible temperature in the universe. In this system, ice melts at approximately 273°K and water boils at approximately 373°K.

Practical Math Problem

25.1 Temperature Conversion

The temperature on a pleasant summer day is 25°C. What is the temperature on the Fahrenheit scale?

> C = temperature in degrees centigrade
> F = temperature in degrees Fahrenheit

$$F = \frac{9}{5}C + 32$$

The temperature is 77°F.

Alternative Forms

$$C = \frac{5}{9}(F - 32)$$

This formula answers questions such as "What is the temperature in degrees centigrade when the temperature is 50°F?" The answer to this question is 10°C.

The formulas for converting centigrade temperatures to kelvin temperatures (and vice versa) are quite simple.

$$C = K + 273.15$$

$$K = C - 273.15$$

Metric to Metric Conversions

Conversions within the metric system are simply a matter of multiplying by the appropriate power of 10, which is one of the advantages of using the metric system.

Length

1 kilometer = 1,000 meters

1 meter = 100 centimeters

1 centimeter = 10 millimeters

Area

1 sq kilometer = 1,000,000 square meters

1 sq meter = 10,000 sq centimeters

1 sq centimeter = 100 sq millimeters

Volume

1 cu kilometer = 1,000,000,000 cu meters

1 cu meter = 1000 liters = 1,000,000 cu centimeters

1 liter = 1,000 cu centimeters = 1,000 milliliters

1 cu centimeter = 1 milliliter = 1,000 cu millimeters

Weight

1 kilogram = 1,000 grams

1 gram = 1,000 milligrams

Tables of Conversion Factors

This section contains the most commonly used conversion factors. Conversion is performed by multiplication, just like converting currency. If $1 is worth 0.72 euros, $5 are worth 5 x 0.72 = 3.60 euros. If one kilogram is "worth" 2.2 pounds, 4 kilograms are "worth" 4 x 2.2 = 8.8 pounds.

Basic Conversions: Length

Convert	To	Multiply By
Miles	Kilometers	1.609
Miles	Yards	1,760
Miles	Feet	5,28
Yards	Feet	3
Yards	Meters	0.914
Feet	Inches	12
Feet	Meters	0.305
Inches	Centimeters	2.54
Kilometers	Miles	0.621
Kilometers	Yards	1,093.6
Kilometers	Feet	3,280.8
Meters	Yards	1.093
Meters	Feet	3.281
Meters	Inches	39.37
Centimeters	Inches	0.394

Basic Conversions: Area

Convert	To	Multiply By
Square Miles	Square Kilometers	2.59
Square Miles	Acres	640
Acres	Square Yards	4,840
Acres	Square Meters	4,046.86
Square Yards	Square Feet	9
Square Yards	Square Meters	0.836
Square Feet	Square Inches	144
Square Feet	Square Meters	0.092
Square Inches	Square Centimeters	6.452
Square Kilometers	Square Miles	0.386
Square Kilometers	Acres	247.10
Square Meters	Square Yards	1.196
Square Meters	Square Feet	10.764
Square Meters	Square Inches	1,550
Square Centimeters	Square Inches	0.155

Basic Conversions: Volume

Convert	To	Multiply By
Cubic Miles	Cubic Kilometers	4.168
Cubic Yards	Cubic Feet	27
Cubic Yards	Cubic Meters	0.765
Cubic Yards	Liters	764.55
Cubic Yards	Gallons	201.97
Cubic Feet	Cubic Meters	0.765
Cubic Feet	Liters	28.317
Cubic Feet	Gallons	7.481
Cubic Feet	Cubic. Inches	1,728
Gallons	Liters	3.785
Cubic Inches	Cubic Centimeters	16.387
Cubic Kilometers	Cubic Miles	0.24
Cubic. Meters	Cubic. Yards	1.308
Cubic. Meters	Cubic. Feet	35.315
Cubic Meters	Gallons	264.17
Liters	Gallons	0.264
Liters	Cubic Feet	0.035
Cubic Centimeters	Cubic Inches	0.061

Basic Conversions: Weight

Convert	To	Multiply By
Tons	Pounds	2,000
Long Tons	Pounds	2,240
Metric Tons	Pounds	2,204.6
Tons	Kilograms	907.2
Long Tons	Kilograms	1,106
Metric Tons	Kilograms	1,000
Kilograms	Pounds	2.205
Kilograms	Ounces	35.28
Pounds	Ounces	16
Pounds	Grams	453.
Ounces	Grams	28.3
Grams	Ounces	0.03
Grams	Grains	15

Basic Conversions: Kitchen

Convert	To	Multiply By
Gallons	Quarts	4
Gallons	Pints	8
Gallons	Fluid Ounces	128
Quarts	Pints	2
Quarts	Fluid Ounces	32
Pints	Fluid Ounces	16
Pints	Cups	2
Cups	Fluid Ounces	8
Cups	Tablespoons	16
Cups	Teaspoons	48
Fluid Ounces	Tablespoons	2
Fluid Ounces	Teaspoons	6
Tablespoons	Teaspoons	3
Teaspoons	Drops	96

Basic Conversions: Energy Derived from Common Fuels

Convert	To	Multiply By
Tons coal	Gallons gasoline	201.8
Tons coal	Gallons LPG	263.8
Tons coal	Gallons #2 Fuel Oil	181.4
Tons coal	Kilowatt-Hours	7,384
Gallons gasoline	Gallons LPG	1.308
Gallons gasoline	Gallons #2 Fuel Oil	0.899
Gallons gasoline	Kilowatt-Hours	36.6
Gallons gasoline	Pounds Coal	9.91
Gallons LPG	Gallons gasoline	0.765
Gallons LPG	Gallons #2 Fuel Oil	0.688
Gallons LPG	Kilowatt-Hours	27.99
Gallons LPG	Pounds Coal	7.58
Gallons #2 Fuel Oil	Gallons gasoline	1.112
Gallons #2 Fuel Oil	Gallons LPG	1.454
Gallons #2 Fuel Oil	Kilowatt-Hours	40.7
Gallons #2 Fuel Oil	Pounds Coal	11.02
Kilowatt-Hours	Gallons gasoline	0.027
Kilowatt-Hours	Gallons #2 Fuel Oil	0.025
Kilowatt-Hours	Gallons LPG	0.036
Kilowatt-Hours	Pounds Coal	0.271

Basic Conversions: Heat and Energy

Convert	To	Multiply By
Watt-Hours	BTUs	3.412
Watt-Hours	Calories	860
Watt-Hours	Calories	0.86
Watt-Hours	Joules	3,600

A calorie is a nutritional unit with 1 calorie equal to 1,000 calories. You need 1 calorie of heat to raise 1 kilogram of water 1 degree centigrade at atmospheric pressure.

Basic Conversions: Power

Convert	To	Multiply By
BTU/Hour	Calorie/Second	0.07
BTU/Hour	Horsepower	0.000 393
BTU/Hour	Joule/Minute	17.584
BTU/Hour	Kilowatt	0.000 293
Calorie/Second	BTU/Hour	14.286
Calorie/Second	Horsepower	0.005 6
Calorie/Second	Joule/Minute	251.2
Calorie/Second	Kilowatt	0.004 187
Horsepower	BTU/Hour	2,545
Horsepower	Calorie/Second	178.18
Horsepower	Joule/Minute	44,760
Horsepower	Kilowatt	0.746

Basic Conversions: Power

Convert	To	Multiply By
Joule/Minute	BTU/Hour	0.0569
Joule/Minute	Calories/Second	0.003 98
Joule/Minute	Horsepower	0.000 022 3
Joule/Minute	Kilowatt	0.000 016 7
Kilowatt	BTU/Hour	3,412
Kilowatt	Calories/Second	238.85
Kilowatt	Horsepower	1.3405
Kilowatt	Joule/Minute	60,000

Basic Conversions: Pressure

Atmosphere	Pascal	101,325
Atmosphere	Pounds/Square Inch	14.697
Atmosphere	Torr	760
Pascal	Atmosphere	0.000 009 9
Pascal	Bar	0.000 01
Pascal	Pounds/Square Inch	0.000 145
Pascal	Torr	0.007 5
Pounds/Square Inch	Atmosphere	0.068
Pounds/Square Inch	Pascal	6,894.76
Pounds/Square Inch	Torr	51.715
Torr	Atmosphere	0.001 316
Torr	Pascal	133.322
Torr	Pounds/Square Inch	0.019 337

1 bar = 100,000 pascals

1 torr = 1 millimeter of mercury

Basic Conversions: Volume Densities

Grams per cc	Kilograms per Liter	1
Grams per cc	Pounds per Cubic Foot	62.43
Grams per cc	Pounds per Cubic Inch	0.036
Grams per cc	Ounce per Cubic Inch	0.578
Pounds per Cubic Foot	Grams per cc	0.016
Pounds per Cubic Foot	Pounds per Cubic Inch	0.000 578
Pounds per Cubic Foot	Ounce. per Cubic. Inch	0.009 3
Pounds per Cubic. Inch	Grams per cc	27.68
Pounds per Cubic Inch	Pounds per Cubic Foot	1,728
Pounds per Cubic Inch	Ounce. per Cubic Inch	16
Ounce per Cubic Inch	Grams per cc	1.73
Ounce per Cubic Inch	Pounds per Cubic Foot	108
Ounce per Cubic Inch	Pounds per Cubic Inch	0.062 5

CHAPTER 26

 Index to All Problems

In This Chapter

➤ Index to All Problems

Chapter 4 Buying and Selling: Percentages

4.1 Buying in Bulk

A small box of cornflakes contains 12 ounces and costs $2.00. A giant box of cornflakes contains 18 ounces and costs $2.50. How much do you save by buying the giant box?

4.2 Choosing a Deductible

An automobile insurance policy with a $100 deductible has a premium of $240. The same policy with a $300 deductible has a premium of $200. What percentage of the time must a claim be filed in order to make the policy with the higher premium a better buy?

4.3 Choosing a Commission Plan

A salesman is offered a choice of two plans: a base salary of $25,000 plus 5% commission on annual sales, or a base salary of $20,000 plus 7% commission on annual sales. How much must he sell annually in order to make more money by accepting the $20,000 plus 7% commission arrangement?

4.4 Reaching a Sales Goal

A salesman makes 5% commission on all his sales. He has made $800,000 in sales through the end of July. How much must he average in monthly sales in order to make $100,000 in commissions for the year?

4.5 Best Use of Two Ingredients

A company manufactures both a high-protein and a high-carbohydrate trail mix. One pound of the high-protein trail mix uses ¾ of a pound of nuts and ¼ of a pound of fruit, and sells for $4. One pound of the high-carbohydrate trail mix uses 5/8 of a pound of fruit and 3/8 of a pound of nuts, and sells for $3. The company has 300 pounds of nuts and 240 pounds of fruit on hand. How many pounds of each type of trail mix should it make to maximize its revenue?

4.5 Best Use of Two Ingredients in a Different Guise

A TV company makes dramas and reality shows, and has 120 actors available to appear in them. A drama costs $4,000,000 to produce and uses four actors, while a reality show costs $1,000,000 to produce and uses six actors. If a drama attracts 3,000,000 viewers and a reality show attracts 2,000,000 viewers, how should the company allocate its budget of $80,000,000 to attract the largest number of viewers? Each actor can only appear in one show.

4.6 Retail Markup Percentage

The retail price of a digital camera is $90. The cost to the retailer is $55. What is the markup percentage?

4.7 Markdown Percentage

A cell phone is marked down from $70 to $50. What is the markdown percentage?

4.8 Two Successive Percentage Increases

The price of a stock appreciates by 10% 1 year and by 20% the next year. By what percentage has the price increased during the 2-year period?

4.9 Two Successive Percentage Reductions

A dress is discounted 30%, and then is discounted an additional 20%. What is the combined discount percentage from the original price?

4.10 Percentage Gain Needed to Recoup Loss

An investment has declined in value by 20%. What percentage gain is needed in the investment in order for the investment to recover to its original value?

4.11 Percentage Loss to Fall Back to Even

An investment has increased in value by 20%. What percentage loss is required in order for the investment to decline to its original purchase price?

Chapter 5 Borrowing Money and Payment Plans

5.1 Simple Interest

How much interest will accrue to a loan of $2,000 earning 3% simple interest for a period of 4 years? What will be the total amount due to the lender at the end of the 4 years?

5.2 Periodic Compound Interest

How much interest will accrue to a loan of $2,000 earning 3% interest compounded quarterly for a period of 5 years? What will be the total amount to be paid to the lender at the end of the 5 years?

5.3 Different Compounding Frequencies

What interest rate compounded semi-annually is equivalent to a 2.2% rate compounded monthly?

5.4 Continuous Compounding

How much interest will accrue to a loan of $2,000 earning 2.5% interest compounded continuously for a period of 8 years? What will be the total amount to be paid to the lender at the end of the 8 years?

5.5 Loan-to-Value Ratio

A bank approves a loan of $450,000 at a loan-to-value ratio of 90%. What is the maximum appraised value of a property that can be purchased with this loan?

5.6 Periodic Payments

What is the monthly payment for a car that costs $12,000 if a 10% down payment is made and the balance is to be paid in monthly installments over a 5-year period with money borrowed at 6%, and the first payment is due a month after the car is purchased?

5.7 Balance Remaining on a Loan

What is the balance remaining on a loan of $20,000 if the money was borrowed at an annual rate of 5% and 24 monthly payments of $500 have been made?

5.8 Funding a Perpetuity

How much is needed to fund a perpetuity that will pay $10,000 annually if the money is invested at 2.5% compounded annually?

5.9 How Long Will an Inheritance Last?

How long will an inheritance of $100,000 last if it is invested at 3% and $7,500 is withdrawn as soon as the inheritance is received and at the beginning of every year thereafter?

Chapter 6 Stocks, Mutual Funds, Bonds, and Options

6.1 Real Rate of Return

Between October 2009 and October 2010, the rate of inflation was 1.18%. During that period, the price of gold increased by 17.39%. What was the real rate of return on gold during that period?

6.2 Return on Investment Annualized

A stock is purchased for $101 and sold 60 days later for a price of $104.75. The stock also paid a dividend of 25¢ during that period. The commissions for buying and selling amounted to 48¢ per share. What is the ROIA for this transaction?

6.3 Dividend Yield Ratio

A stock that is currently trading at $80 per share pays quarterly dividends of 48¢. What is the dividend yield ratio?

6.4 Earnings per Share

A stock has 40,000,000 outstanding common shares (*outstanding* is a term meaning the number of shares issued by the company). The company has annual earnings of $2,500,000. What are this stock's earnings per share (abbreviated EPS)?

6.5 Price/Earnings Ratio

A stock is currently trading at $40 per share and has earnings of $2.40 per share. What is its price/earnings ratio (abbreviated P/E)?

6.6 Margin Call Trigger Price

A stock selling at $80 is purchased for an initial payment of 70%. If the maintenance margin is 55%, at what price will a margin call be triggered?

6.7 Yield to Maturity

A bond has 5 years until it matures at a face value of $1,000. It pays an annual coupon of $40 and is currently priced at $950. What is the yield to maturity?

6.8 Equivalent Taxable Yield

A bond has a tax-free yield of 1.45%. What is the equivalent taxable yield for a person with a marginal tax rate of 28% (a.k.a. an income tax bracket of 28%)?

6.9 ROIA on Covered Calls

IBM is currently trading at $164. You sell an IBM April 170 call for $1.91 a share; the actual price is $191 as the option is for 100 shares of stock. There are 60 days until expiration. What is the annualized return on the option if you still own the shares at expiration?

Chapter 7 Taxes and Other Governmental Math

7.1 Tax-Deductible Contributions

What is the true cost of a $100 contribution to a tax-deductible charity for an individual in the marginal 28% tax bracket?

7.2 Net Monthly Payments

What is the net monthly payment after 120 monthly payments of $2,147.29 have been made on a 30-year loan of $400,000 at 5% for a person who is in the marginal 28% tax bracket? The net monthly payment is the payment less the amount saved due to the deductibility of interest.

7.3 Quarterly Tax Payments

In one scenario, you pay quarterly estimated payments of $6,000. In the other scenario, you pay a lump sum at the end of the year at a penalty of $100. The money you would have used to pay quarterly taxes could instead be put in a CD that earns 2% annually. Should you make the quarterly payments or put the money in a CD and pay a lump sum at the end of the year?

7.4 Straight-Line Depreciation

What is the annual straight-line depreciation on a computer purchased for $600 with an expected life of 4 years and a salvage value of $80? What is the book value at the end of the 3rd year?

7.5 Double Declining Balance Depreciation

A truck is purchased for $20,000. It has an expected life of 10 years and a salvage value of $3,000. What is the book value at the end of the 4th year, and what is the depreciation for the 4th year, and as computed by the double declining balance method?

7.6 Sum of the Years Digits Depreciation

A drill press is purchased for $6,000. It has an expected life of 8 years. What is the book value at the end of the 5th year, and what is the depreciation for the 5th year, as computed by the sum of the years digits method?

7.7 Social Security

If you retire at age 62, you will receive a monthly Social Security check of $1,200. If you retire at age 66½, you will receive a monthly Social Security check of $1,500. What age will you be when the total income from retirement at 66½ exceeds the total income from retirement at 62?

7.8 Keeping Up with Inflation

Your after-tax income increased from $30,000 in 2009 to $30,800 in 2010. Did this keep up with inflation?

Chapter 8 Everyday Problems

8.1 Changing the Size of a Recipe

A recipe for meat loaf calls for 3 pounds of hamburger and serves 8. How much hamburger will be needed to serve 12?

8.2 Reading a Map

The scale of a map is 1 inch to 50 miles. What does a distance of 5 inches on the map represent?

8.3 Time Needed for a Two-Person Job

John can wash the dishes in 30 minutes if he does them by himself. Sue can wash the dishes in 20 minutes if she does them by herself. How long will it take both of them working together to wash the dishes?

8.4 Time Needed to Address Christmas Cards

If three people can address 20 Christmas cards in 50 minutes, how long will it take 4 people to address 35 Christmas cards?

8.5 Mixing Two Mixtures

A 2-quart pitcher of fruit punch is 60% fruit juice. How many quarts of fruit punch that is 80% fruit juice must be mixed with this in order to have a mixture that is 75% fruit juice?

8.6 When Will the Soup Cool?

After it has been placed in the microwave for heating, the temperature of a bowl of soup is 160°. It takes 5 minutes for it to cool to 140°. How long will it take to cool to 120° if the temperature of the room is 70°?

8.7 The Cost of Electricity

Electricity costs 12¢ per kilowatt-hour, and a typical computer and monitor use about 0.3 kilowatts per hour. How much does it cost to leave the computer and monitor on for 8 hours nightly?

8.8 Gasoline Savings from Driving a Hybrid

A car with a standard internal combustion engine gets 30 miles per gallon, whereas a hybrid gets 50 miles per gallon. Gasoline costs $3.00 a gallon, and you drive an average of 12,000 miles annually. What is your annual savings on gasoline from buying a hybrid?

8.9 Driving to Buy at a Cheaper Price

Gasoline costs $3.00 a gallon and your car gets 25 miles a gallon. How far would you go to save $30 on the purchase of a digital camera?

Chapter 9 Automobile Usage and Performance

9.1 Savings from Better Mileage

What is the annual savings from driving 12,000 miles per year with a car that gets 12 miles per gallon as opposed to a car that gets 10 miles per gallon, if fuel costs $3 per gallon?

9.2 Savings from Fuel Additives

You own a car that currently gets 24 miles per gallon. Gas costs $3.00 per gallon. A fuel additive costs $2.00 per ounce, and one ounce must be added to 8 gallons of gasoline to be effective. What mileage must the car get with the fuel additive in order to justify the cost of the additive?

9.3 Bore, Stroke, and Engine Displacement

What is the displacement of a six-cylinder engine with a 4-inch bore and a 3.5-inch stroke?

9.4 Compression Ratio

An engine has a 4-inch bore, a 3.5-inch stroke, a measured chamber volume of 4.27 cubic inches, and a gasket thickness of 0.05 inches. What is its compression ratio?

9.5 Milling the Heads

By how much must the cylinder heads be milled on an engine with a 3.5-inch stroke in order to increase the compression ratio from 9.6 to 10.5?

9.6 Horsepower of an Auto Engine

What horsepower is delivered by a six-cylinder engine in which a single cylinder has a radius of 3 inches, a stroke of 8 inches, a crank that makes 250 revolutions per minute, and a mean effective pressure of 100 pounds per square inch?

9.7 Checking Your Speedometer

It takes you 87 seconds to complete a measured mile during which your speedometer registered a constant 40 miles per hour. What is your true speed, and what percentage of your measured speed is your true speed?

9.8 Tire Diameter Effect on the Speedometer

If the tire diameter is changed from 36 inches to 38 inches, what is the true speed when the speedometer reads 60 miles per hour?

9.9 Force Needed to Hold Car on Hill

A 4,000-pound car is parked on a hill at a 9% angle. How much force is required to prevent the car from rolling down the hill?

9.10 Maximum Acceleration in Towing a Car

A rope will break if stretched with a force of 3,000 newtons or more. What is the maximum acceleration with which it can tow a 1,500-kilogram car on level ground?

9.11 Distance Traveled After Braking

A car traveling 50 miles per hour brakes suddenly and comes to a complete stop in 6 seconds. How far does it travel between the time the brakes are applied and the time it comes to a complete stop?

Chapter 10 The World Around Us

10.1 Time Saved by Going Faster

How much time can you save on a 200-mile trip by averaging 60 miles per hour rather than 50 miles per hour?

10.2 Making Up for Lost Time

You wish to average 50 miles per hour on a 300-mile trip, but unfortunately you only averaged 40 miles per hour for the first 2 hours. How many miles per hour must you average on the remainder of the trip to make up for lost time?

10.3 Point of No Return

The distance from Los Angeles to Tokyo is 5,500 miles. A jet can fly 600 miles per hour in still air, but there is a west-to-east headwind of 50 miles per hour. How long after takeoff from Los Angeles will it be until the plane reaches the point of no return, where it takes the same amount of time to continue to Tokyo as it does to return to Los Angeles?

10.4 How Far Will Your Fuel Take You?

A boat with a 12-gallon tank capacity gets 30 miles per gallon inbound (with the tide) and 20 miles per gallon outbound (against the tide). How far out can it safely venture?

10.5 Determining Air Speed and Wind Speed

A plane traveling west from Denver makes the 1,000-mile trip to Los Angeles in 2 hours. Going east 1,800 miles to Orlando takes 3 hours. Assuming that the wind conditions are the same in both directions, what is the speed of the plane and what is the wind speed?

10.6 Crossing a River Against the Current

A motorboat is capable of going 15 miles per hour in still water. A current of 4 miles per hour is flowing in a river that is ¼ of a mile wide. At what angle should the boat travel in order to land at the point on the opposite bank directly across from where it starts, and how long will the trip take?

10.7 Cost of Currency Conversion

Suppose that the currency conversion rate is quoted as above: dollars to euros 0.72-0.75. What percentage is lost by converting dollars to euros and back again?

10.8 Height Measurement Using One Angle

What is the height of a tree that is 50 feet away from an observer, and the angle of elevation from the observer to the top of the tree is 62°?

10.9 Height Measurement Using Two Angles

The angle of elevation to the top of a tree on the other side of a river is 40°. If you walk 100 feet in a straight line directly away from the tree, the new angle of elevation is 28°. How wide is the river? How tall is the tree? (As mentioned, this method is useful when one cannot directly measure the distance to the object, as in the preceding example.)

10.10 Distance by Triangulation

Two observers on a beach are separated by 3 miles. They see a boat at sea between them, and measure the angles between the line to the boat and the line between the two observers. One observer measures an angle of 35° and the other an angle of 61°. How far is the boat from the shore?

Chapter 11 *To Your Good Health*

11.1 The Body Mass Index (BMI)

What is the Body Mass Index of a man who is 5 feet 9 inches tall and weighs 150 pounds?

11.2 A Schedule for a Diet

An overweight individual wants to go on a diet to go from his current weight of 220 pounds to 180 pounds within three months. How many pounds should he plan on trying to lose weekly?

11.3 Maximum Heart Rate

What is the maximum heart rate for a typical 55-year-old?

11.4 Time until Sunburn

What SPF factor sun block is needed to enable an individual to stay in the sun for 3 hours, if he normally experiences sunburn in 10 minutes without protection?

11.5 Blood Alcohol Content

How long should a 150-pound female wait after consuming 6 ounces of 70 proof alcohol to make sure that her blood alcohol content is below 0.08?

11.6 Saving on Dietary Supplements

A patient needs at least 100 units of iron and 120 units of niacin daily. Two supplements are available. The first supplement contains 5 units of iron and 20 units of niacin per ounce, and costs 20¢. The second supplement contains 10 units of iron and 5 units of niacin per ounce, and costs 30¢. What is the cheapest combination of the two supplements that will satisfy the daily nutritional needs, and how much of each supplement do you need?

11.6 Saving on Dietary Supplements in a Different Guise

A garage owner hires two mechanics, Al and Bob, on a daily basis. In a typical day, Al could repair 3 cars and 3 trucks. Al's daily salary is $190. Bob can repair 2 cars and 4 trucks per day and has a daily salary of $220. If there are 12 cars and 18 trucks that need repairs, how many days of each worker should the owner hire?

11.7 Correct Dosage

A patient needs 40 milligrams (mg) of a medication intramuscularly. The drug is available as 100 mg/5 ml. How many milliliters (ml) of the drug should be administered?

11.8 Drug Dosage by Drips

500 ml of a drug are to be delivered intravenously over a 4-hour period. The tubing drip factor is 10 drips per milliliter (Note: *gtt* is a standard abbreviation for a single drip, should you ever see this notation). How many drips per minute should be administered?

Chapter 12 Recreation

12.1a Batting Averages

What is the batting average of a player who has had 350 official at bats and 87 hits?

12.1b Slugging Percentage

What is the slugging percentage of a player who had 27 home runs, 4 triples, 36 doubles, and 71 singles in 477 official at bats?

12.1c Earned Run Average

What is the earned run average of a pitcher who allowed 23 earned runs in 85 2/3 innings pitched?

Earned run average = 9 × number of earned runs allowed / number of innings pitched

12.2 NCAA Quarterback Efficiency Rating

A college quarterback threw 25 passes in a game, completing 14 of them for 205 yards and 1 touchdown. Two passes were intercepted. What is his efficiency rating?

12.3 Probability of Making Free Throws

A player who makes 80% of his free throws is shooting 3 free throws. What is the probability that he will make exactly 2 free throws?

12.4 Final Velocity of a Dragster

From a standing start, a drag racer covers a ¼ mile in 11 seconds. Assuming that his acceleration is constant, what is his velocity at the end of the ¼ mile?

12.5 Probability of Winning by 2 Points

You have reached a moment in a game where the first person to score 2 more points than his opponent wins (such as 10 to 10 in ping-pong or deuce at tennis). If your probability of winning each point is 60%, what is your probability of winning the game?

12.6 Computing Your Golf Handicap

You recently took a golfing vacation to Farmingdale, Long Island, where you played five rounds of golf on the notorious Bethpage Black, with a course rating of 75.4 and a slope factor of 144. Your scores were 97 (you got lucky), 108 (lots of rough), 99 (actually it was 100, but your club really didn't touch the ball that time, so you didn't count that stroke), 102, and 106. What is your handicap as determined by these five rounds?

12.7 Observing All Species of Birds

A local bird-watching venue has 15 different species of birds, and the probability of seeing a bird of any 1 species is approximately the same as seeing a bird of any other species. On average, how many birds will you have to observe before you've seen 1 of each species?

12.8 Observing Several Rare Birds

What is the probability of seeing 3 birds of a particular rare species in the same day if on an average day, you will only see 1?

12.9 Magic Number for a First-Place Team

On February 22, 2011, the San Antonio Spurs were in first place in the Southwest Division of the NBA's Western Conference, with a record of 46 to 10 (46 wins, 10 losses). In second place were the Dallas Mavericks, with a record of 40 to 16. The NBA season is 82 games long. What was the magic number for the Spurs at that time?

12.10 Win Percentage Needed

A baseball team has played 100 of its 162 games and has won 58 of them. All 35 of its remaining games are at home, where it wins 65% of its games. What percentage of the road games must it win in order to have an overall winning percentage of at least 60%?

12.11 Percentage of Uncounted Ballots

With 80% of the vote counted, a candidate has 55% of the vote. What percentage of the remaining ballots must vote for him in order to ensure he gets a majority of the votes?

Chapter 13 Understanding Gaming

13.1 Expected Value of a Bet

You have bet $20 at odds of 5-2 (meaning that if you win, for every $2 you risked, you will win $5) on a team that you feel has a 30% chance of winning. What is the expected value of your bet?

13.2 House Percentage: Two-Sided Bet

We are calling this a two-sided bet simply because you can take either side of the line. Yes, there is a different type of line involving several options; we'll discuss that shortly.

What is the house percentage on a game in which the line is -250 + 210?

13.3a Overcoming the House Percentage

You are betting $1 to win $1.20. What percentage of bets must you win in order to break even?

13.3b Win Percentage of a Random Bet

Suppose that you are betting $1 to win $1.20 in a situation where the house percentage is 4%. What is the win probability of a similarly placed random bet in which someone randomly chose to bet $1 to win $1.20 where the house percentage is 4%?

13.4 House Percentage: Many Contestants

In a 3-horse race, the odds on the 3 horses are 6-5, 2-1, and 5-2. What is the house percentage on this race?

13.5 House Percentage on a Parlay

What is the house percentage on a 2-team parlay at odds of 13-5?

13.6 Changing Horses in Midstream

A bettor has bet $300 to win on the underdog in a tennis match and received 3-1 odds. The underdog wins the first set, goes up a break in the second, and the odds change so that the bettor can now get 2-1 on the original favorite. How much should he bet at those odds to be sure of winning the same amount no matter who wins, and how much will that amount be?

13.7 Hedging with Multiple Contestants

A bettor has bet $200 on the favorite in a four-person event at odds of 2-1. He later can obtain odds of 7-1, 11-1, and 23-1 on the other three entrants. How much should he bet on each to assure himself of the same return no matter who wins?

13.8 Multiple Bets with Equal Returns

A bettor wishes to bet on three teams in such a way that he will receive the same payoff no matter which of the three teams wins. The odds against the three teams are 7-1, 11-1, and 23-1. If he has $120 to bet, how much should he bet on each team, and what odds is he receiving on his money?

13.9 Gambler's Ruin

Assume that you have a stake of $50 and will continually make bets of $1 on a proposition that you feel you have a probability of .51 of winning. What is the probability that you will increase your stake to $125 before you go broke?

Chapter 14 In the Workshop

14.1 Speed of a Driven Gear

A gear with 16 teeth is rotating at 120 rpm. It is driving a gear with 24 teeth. At what speed is the driven gear rotating?

14.2 Speed of a Wheel Connected by a Belt

A wheel with a 2-foot diameter is rotating at 20 rpm. It is linked by a belt to a wheel with an 8-inch diameter. At what speed is the driven wheel rotating if there is no slippage?

14.3 Drill Feed

The drill feed is the distance that the drill advances in a single revolution. It is usually measured in inches per revolution.

What is the feed of a drill that rotates at 300 rpm and drills a hole 2 inches deep in 20 seconds?

14.4 Required Drill Depth

How deep should you feed a 74° countersink to drill a hole 0.65 inches in diameter?

14.5 Cutting Speed of a Mill or Lathe

How many revolutions per minute must a cutting mill with a 2-inch diameter make in order to cut machine steel at 80 surface feet per minute?

14.6 Polar Moment of Inertia (Hollow Rod)

A hollow rod has an inner radius of 3 centimeters and an outer radius of 5 centimeters. What is its polar moment of inertia?

14.7 Maximum Torque for a Cylindrical Rod

For the hollow rod specified in problem 14.6, what is the maximum possible torque that can be applied to the rod if the shear stress cannot exceed 50 megapascals?

14.8 Twist Angle for a Cylindrical Rod

Copper has a shear modulus of elasticity of 936,000,000 pounds per square foot. What is the twist angle in radians of a hollow copper rod 2 feet long with an inner radius of 1 inch and an outer radius of 2 inches when a torque of 5,000 foot-pounds is applied?

14.9 Lateral Strain of a Compressed Bar

A steel bar with a rectangular cross-sectional area of 40 square inches is subject to an axial compression of 200,000 pounds. Steel has a modulus of elasticity of 2.8×10^7 pounds per square inch and a Poisson ratio of -0.29. What is its lateral strain?

14.10 Change in Length of a Bar

The modulus of elasticity (a.k.a. Young's modulus) of aluminum is 10,000,000 pounds per square inch. What is the change in length of an aluminum bar 40 inches long with a cross-sectional area of 25 square inches when subjected to a force of 6,000 pounds?

14.11 Cylindrical Rod Resting in a Slot

A rod of radius 8 inches rests touching two sides of a slot. The width of the slot at the bottom is 3 inches, and the two sides make angles of 34° and 52°. How far above the floor of the slot is the lowest point on the rod?

Chapter 15 Moving Stuff from Place to Place

15.1 Using a Lever

How much force must be applied to a lever to lift a 2,000-pound weight if the lever is 6 feet long and the fulcrum is positioned 6 inches from the end of the lever closest to the weight?

15.2a Using a Frictionless Inclined Plane

How much force is needed to push a 1,000-pound weight up a frictionless inclined plane 10-feet long to a height 2 feet above the ground?

15.3 Wheel and Axle

A wheel with a radius of 2 feet is concentric with an axle of a radius of 3 inches. A cord is wrapped around the axle and fastened to a 100-pound weight. How much force must be applied to the wheel to lift the weight, and how many revolutions must the wheel turn to lift the weight 3 feet?

15.4 Forces on a Boom and Cable

A boom of negligible weight makes an angle of 50° with the ground. A cable passes horizontally from a wall to the boom and the cable then suspends vertically to support a weight of 500 pounds. What is the force on the boom and the tension in the cable?

15.5 Rolling a Cylindrical Drum Up a Step

How much force does it take to roll a horizontal cylinder weighing 200 pounds with a radius of 3 feet up a step that is 2 feet high by pushing at the top of the cylinder parallel to the ground?

15.6 Fluid Speed in a Pipe

The cross-sectional area of a pipe at point A is 5 square inches, and the cross-sectional area of the same pipe at point B is 4 square inches. The speed of the fluid flow at point A is 7 feet/second. What is the speed of the fluid flow at point B?

15.7 Power Output from Fluid Pressure Drop

A flow rate of 50 gallons per minute produces a pressure drop of 20 pounds per square inch across a turbine. What is the power output of the turbine in horsepower?

15.8 Flow Rate of a Pump

A 2-horsepower pump with an efficiency of 80% produces an increase in water pressure of 10 pounds per square inch. What flow rate is necessary to produce this?

15.9 Converting Pressure to Head

Ethyl alcohol has a specific gravity 0.789 and is flowing through a pipe under a pressure of 40 pounds per square inch. What is the head in feet?

15.10 Velocity of Water Flow in a Pipe

What is the speed at which water under a head of 80 feet is discharged from a pipe that is 4,000 feet long and 2 feet in diameter?

Chapter 16 Engines and Motors

16.1a Engine Power

A gasoline engine burns 3 gallons of gasoline per hour. Gasoline has a heating value of 20,000 BTU/pound and a specific gravity of 0.8. What is the input power of the engine in kilowatts? (1 BTU is one British thermal unit, the amount of heat needed to raise 1 pound of water 1 degree Fahrenheit at 1 atmosphere pressure).

16.1b Engine Efficiency

The engine in problem 16.1a produces an output power of 30 kilowatts. What is its efficiency?

16.2 Limit of Engine Efficiency

What is the maximum possible efficiency of an engine in which heat enters the engine at 1,200°C (centigrade) and exits the engine at 300°C?

16.3a Energy Consumption and Engine Power

What is the energy used in kilowatt-hours by a 3-horsepower compressor that runs for 2 hours?

16.3b Cost of Running an Engine

What is the financial cost of running the compressor in problem 16.3a if electricity costs 12¢ per kilowatt-hour?

16.4 Horsepower Needed to Do Work

What horsepower is delivered by an engine that does 1,000 foot-pounds of work per second?

16.5 Horsepower Transmitted by a Belt

What is the horsepower transmitted by a belt 12 inches wide with an allowed pull of 80 pounds per inch of width and a belt speed of 4,000 feet per minute?

16.6 Lifting Velocity of a Motor

A 10-kilowatt engine has an efficiency of 85%. With what constant speed can it lift a 1,000-pound weight?

16.7 Smallest Rotor Diameter for Motor

What is the smallest diameter for a rotor shaft for a 10-horsepower motor that will rotate at 6,000 rpm and will be subject to a maximum shear stress of 5,000 pounds per square inch?

16.8 Wall Thickness to Withstand Pressure

A thin-walled cylinder with a 30-inch diameter is to be made of steel, which has a yield stress of 30,000 pounds per square inch. It must withstand pressures of 500 pounds per square inch and have a safety factor of 3. How thick must the cylinder wall be?

Chapter 17 Work Needed for Specific Problems

17.1 Heating Water

It requires 4.18 joules to raise the temperature of 1 gram of water 1°C. How much heat is required to raise 500 grams of water from a temperature of 40°C to 80°C?

17.2 Rate of Temperature Change from Heating

Water has a specific heat of 4.18 joules per gram per degree centigrade. If 500 grams of water are heated with a 3-kilowatt heater, at what rate will the water temperature rise?

17.3 Lifting a Weight with a Cable

An 800-pound weight is attached to the bottom of a cable that weighs 4 pounds per foot and goes over the edge of a building. If the weight is initially 200 feet from the top of the building, how much work is done pulling the cable and weight up 10 feet?

17.4 Work Done in Reversible Compression

If 100 grams of nitrogen are reversibly compressed from 300 cubic centimeters to 100 cubic centimeters at a constant temperature of 25°C, how much work is done in the process?

17.5 Partially Filled Rectangular Boxes

A rectangular tank with a length of 8 feet, a width of 6 feet, and a height of 5 feet is filled to a depth of 3 feet with water, which weighs 62.4 pounds per cubic foot. How much work is done pumping out the tank?

17.6 Partially Filled Cylinders

A cylindrical tank has a radius of 3 feet and a height of 7 feet. It is filled to a depth of 5 feet with water, which has a density of 62.4 pounds per cubic foot. How much work is done pumping the water out of the tank?

17.7 Partially Filled Triangular Trough

A tank in the shape of a triangular prism has a length of 10 feet. Its front and back ends are each inverted isosceles triangles with a height of 5 feet and a base of 3 feet. It is filled with water, which has a density of 62.4 pounds per cubic foot, to a depth of 2 feet. How much work is done pumping the water out of the tank?

17.8 Partially Filled Inverted Cone

A tank in the shape of an inverted cone with a radius of 4 feet and a height of 10 feet is filled to a depth of 6 feet with water, which has a density of 62.4 pounds per cubic foot. How much work is done pumping the water out of the tank?

17.9 Partially Filled Hemisphere

A hemispherical tank rests on its base, which has a radius of 5 feet. It is filled to a depth of 3 feet with water, which has a density of 62.4 pounds per cubic foot. How much work is done pumping the water out of the top of the tank?

Chapter 18 Physics

18.1 Weight of Material

What is the weight of 12 cubic feet of water?

18.2 Fluid Pressure at a Given Depth

What is the pressure at the bottom of a swimming pool that is 9 feet deep?

18.3 Thermal Expansion

An aluminum bar 36 inches long is heated from 70°F to 120°F. How much does it expand? The coefficient of thermal expansion of aluminum is 1.23×10^{-5} per degree Fahrenheit.

18.4 Heat Loss Through a Surface

The temperature differential on two sides of a wall 10 feet high and 20 feet wide on a cold winter day is 70°F. What is the total amount of heat transferred through the wall across 4 inches of fiberglass insulation in a day? The thermal conductivity of fiberglass is 0.0232 BTUs per foot per degree Fahrenheit per hour.

18.5 Equilibrium Temperature

A 500-gram sheet of copper at 40°C is placed in contact with a 200-gram sheet of aluminum at 20°C. The specific heat of copper is 0.385 joules per gram per degree centigrade, and the specific heat of aluminum is 0.897 in the same units. What is the temperature when they equalize?

18.6 Heat Transferred at Equilibrium

Given the data in problem 18.5, how much heat is transferred from the copper to the iron?

18.7 Isothermal Compression of Ideal Gas

If 100 cubic feet of an ideal gas at 2 atmospheres pressure is compressed isothermally (keeping the temperature constant) to a volume of 40 cubic feet, what is the pressure of the compressed gas?

18.8 Isobaric Compression of Ideal Gas

If 200 cubic feet of an ideal gas at 50°C is compressed isobarically (keeping the pressure constant) to a volume of 50 cubic feet, what is the temperature of the compressed gas?

18.9 Trajectory of a Thrown Ball

From the top of a hill 100 feet above ground level, a ball is thrown upward at an angle of 37° to the ground with an initial velocity of 120 feet per second. What is its maximum elevation, how long is it until the ball strikes ground, and what is the horizontal distance from the point where it was thrown to where it strikes the ground?

18.10 Determining the Depth of a Well

A rock is dropped into a well and the splash is heard 2 seconds later. If the speed of sound is 1,110 feet per second, how deep is the well?

18.11 Relative Intensity of Sound

A shot from a .357 Magnum pistol registers about 165 decibels. How much more intense is this than the sound of a rock-and-roll singer, who registers 140 decibels?

18.12 Relative Intensity of Earthquakes

In January 2010, an earthquake in Haiti killed more than 220,000 people and had a magnitude (Richter reading) of 7.0. The March 2011 earthquake in Japan resulted in fewer than 10,000 fatalities and had a magnitude of 8.9. How much more energy was released in the earthquake in Japan than in the Haiti earthquake?

18.13 Frequency of a Vibrating String

A piano wire is 80 centimeters long and weighs 4 grams. At what tension must it be stretched in order to vibrate at middle C (262 hertz)?

18.14 Radioactive Decay

Technetium-99m (used in some medical tests) has a half-life of 6 hours. The half-life is the time needed for half of a substance to decay to a nonradioactive state. If 80 milligrams of technetium-99m are administered, how much will be left 24 hours later?

Chapter 19 Electricity

19.1 Ohm's Law

What is the potential difference in volts resulting from passing a current of 20 amperes through a resistance of 5 ohms?

19.2 Total Resistance in a Circuit

Suppose that three resistors with resistances of 2, 3, and 6 ohms are connected in series. What is the total resistance of the circuit? What is the total resistance of the circuit if they are connected in parallel?

19.3 Total Capacitance in a Circuit

Suppose that three capacitors with resistances of 10, 15, and 30 farads are connected in series. What is the total capacitance of the circuit? What is the total capacitance of the circuit if they are connected in parallel?

19.4 Total Non-Coupled Inductance

Suppose that three non-coupled inductors with inductances of 6, 9, and 18 henrys are connected in series. *Non-coupled* means that the inducting field of one does not affect any of the others. What is the total inductance of the circuit? What is the total inductance of the circuit if they are connected in parallel?

19.5 Power Dissipation in a Resistor

A 5-amp current passes through a 4-ohm resistor. How much power is dissipated?

19.6 Horsepower of an Electric Motor

What is the horsepower of a 115-volt motor that draws 3 amps of current and operates at an efficiency of 85%?

19.7 Horsepower at Full-Load Torque

What is the horsepower of a motor that runs at 1,800 rpm with a full-load torque of 4 foot-pounds?

19.8 Power Factor (Current and Voltage)

A 21-watt fluorescent lamp ballast draws 21 watts of power off a 120 volt line carrying 0.39 amps. What is the power factor?

19.9 Capacitative Reactance

What is the capacitative reactance of a 5,000 μF (microfarads) capacitor operating on current with a frequency of 200 hertz?

19.10 Inductive Reactance

What is the inductive reactance of a coil with an inductance of 100 μH (microhenrys) operating on current with a frequency of 60 hertz?

19.11 Impedance

What is the total impedance of a circuit that has a 3-ohm resistor, an inductor with an inductive reactance of 5 ohms, and a capacitor with a capacitative reactance of 2 ohms?

19.12 Power Factor (Resistance and Impedance)

A solenoid valve has a resistance of 970 ohms with an impedance of 1,560 ohms. What is the power factor?

19.13 Resonant Frequency

What is the resonant frequency of an LC circuit with an inductance of 100 μH (microhenrys) and a capacitance of 5,000 μF (microfarads)?

19.14 Synchronous Motors

What is the rotation speed of a 2-pole induction motor connected to a 60 hertz power supply?

19.15 Transformer Current and Voltage

A 10-amp current at 110 volts is stepped up to 440 volts by a transformer. What is the output current?

19.16 Transformer Number of Coil Turns

A transformer has 10 turns in its primary coil and 40 turns in its secondary coil. If the incoming voltage is 110 volts, what will be the outgoing voltage?

Chapter 20 The Most Economical Way to Do Things

20.1 Maximum Profit from Sales

It costs $30 to manufacture a digital camera. If the sale price is $70, 1 million cameras can be sold. Research shows that for each $1 increase in the sales price, 20,000 fewer cameras will be sold, and for each $1 decrease in the sales price, 20,000 more cameras will be sold. What sales price for the camera will maximize profits?

20.2 Pricing with a Fixed Number of Items

A convention center has space for 300 display booths. It can rent all the display booths if it charges $200 per booth, but for each extra $5 that it charges, one less booth will be rented. It costs $30 to maintain a booth that has been rented, but only $10 to maintain a booth that has not been rented. At what price should it rent its display booths?

20.3 Optimum Crop Yield

If 20 apple trees are planted in an acre of ground, they will yield an average of 32 bushels per tree. For each extra tree planted per acre, the average yield per tree is reduced by 2 bushels. How many trees should be planted per acre to obtain the greatest total yield per acre, and what is the optimum yield?

20.4 Cheapest Rectangular Box

A rectangular box is to be constructed to contain a volume of 400 cubic inches. Material for the top costs 10¢ per square inch, material for the bottom costs 20¢ per square inch, and material for the sides costs 8¢ per square inch. What are the dimensions of the cheapest such box, and what will it cost?

20.5a Largest Cylindrical Silo

A silo is to be built in the shape of a cylinder surmounted by a hemisphere. The floor of the cylinder costs $10 per square foot, the walls of the cylinder cost $15 per square foot, the surface of the hemisphere costs $25 per foot, and the floor between the cylinder and the hemisphere costs $12 per square foot. A budget of $100,000 is available for construction. What are the dimensions and the volume of the silo with the largest combined volume for cylinder and hemisphere?

20.5b Cheapest Cylindrical Silo

Assume that a silo is to be built with the same parameters as in the previous problem, but that it must have a combined cylinder and hemisphere volume of 20,000 square feet. What are the dimensions of the silo and how much will it cost?

20.6 Smallest Restocking Processing Costs

An electronics store needs to ship 600 big-screen TVs annually from the warehouse to the outlet. It costs $20 to process a single order and $5 to load and ship a single TV. It also costs $15 annually to store a single TV at the outlet. In what size lots should the store reorder and restock TVs in order to minimize total costs?

20.7 Optimal Use of Labor

A pretzel-making machine rents for $100 per day and can make 250 pretzels per hour. It costs $50 an hour to hire an operator to supervise the machines. How many machines should be rented to fill an order for 8,000 pretzels and what is the total cost?

20.8 Cheapest Way to Lay Cable

A bridge crosses a river that is 100 yards wide. Cable must be laid from point A on one side of the river 200 yards from the bridge to the junction B of the bridge with the other side of the river. It costs $800 to lay cable underwater and $500 to lay cable on dry land. The best way is to locate a point C on the other side of the river, lay cable underwater to C, and then along dry land to B. What is the cheapest way to lay the cable, and how much will it cost?

Chapter 21 The Best Way to Do Things

21.1 Largest Rectangle with Fixed Perimeter

What is the largest rectangular field that can be enclosed using 100 feet of fencing, and what are its dimensions?

21.2 Smallest Perimeter of Rectangle with Fixed Area

What is the smallest amount of fence that is needed to enclose a rectangular field whose area is 400 square feet, and what are the dimensions of the field?

21.3 Greatest Revenue from Oval Display

A convention center is to be constructed in the form of a rectangle with a semicircle at each end. The convention center is to have a perimeter of 1 mile equal to 1,760 yards. The rectangle will produce an average revenue of $50 per square yard, whereas the semicircles will generate an average revenue of $30 per square yard. What should be the dimensions of the convention center to generate the most revenue?

21.4a Minimum Perimeter from Oval Racetrack

A racetrack is to be constructed in the form of a rectangle with a semicircle at each end. If the enclosed area is to be 1,000 square yards, what dimensions of the track will produce the minimum perimeter, and what is that perimeter?

The picture associated with this problem is the one in problem 21.3.

21.4b Minimum Perimeter from Oval Racetrack

Suppose that in problem 21.4a we require that the area of the rectangle is to be 1,000 square yards. What dimensions of the track will produce the minimum perimeter and what is that perimeter?

21.5a Smallest Poster with Fixed Print Area

What are the dimensions of the rectangular poster with the smallest area that has side margins of 1 inch, a top margins of 2 inches, a bottom margin of 1.5 inches, and is to contain a rectangular printed area of 60 square inches?

21.5b Largest Poster Area Inside Margins

A rectangular poster with an area of 300 square inches is to have a top margin of 3 inches, a bottom margin of 2 inches, and side margins of 1 inch each. What are the dimensions of the poster that will yield the largest area inside the margins?

21.6 Where to Stand to View a Painting

A picture 8 feet high is hung so that the bottom of the picture is 2 feet above eye level. At what distance should an observer stand in order to maximize the angle subtended by the picture?

21.7 Carrying a Pipe Around a Corner

What is the longest pipe that can be carried around the corner of an L-shaped corridor that has widths of 6 and 8 feet and is 10 feet high?

21.8 Shortest Ladder Going Over a Wall

A wall 10 feet high is 16 feet away from a building. What is the shortest ladder that will reach from the ground over the wall to the building?

21.9a Strongest Beam Cut from a Log

What are the dimensions of the strongest rectangular beam that can be cut from a cylindrical log with a diameter of 2 feet?

21.9b Stiffest Beam Cut from a Log

What are the dimensions of the stiffest rectangular beam that can be cut from a cylindrical log with a diameter of 2 feet?

21.10a Most Light Through a Norman Window

A Norman window is a rectangle surmounted by a semicircle. If the glass used for the semicircle transmits 10 lumens per square foot and the glass used for the rectangle transmits 20 lumens per square foot, what are the dimensions of a Norman window with a perimeter of 25 feet that transmits the most light?

21.10b Smallest Perimeter of Norman Window

If the total enclosed area of the Norman window of 21.10a is to be 40 square feet, what are the dimensions of the Norman window with the smallest perimeter, and what is the perimeter?

Chapter 22 Mathematical Tools for Business Management

Most of the tools in this chapter are formulas or ratios for evaluating various aspects of the business.

22.1 Business Break-Even Point

The fixed costs for a GPS manufacturer are $3,000,000 per year. It can produce GPS devices for $40, which it sells for $100. Assuming that it sells every GPS it produces, what is its break-even point?

22.2 Current, Quick, and Cash Ratios

22.3 Cash Flow from Operations Percentage

Chapter 23 Plane Geometry

23.1 Area and Perimeter of a Rectangle

A rectangle is 12 feet wide and 18 feet long. What are its perimeter and area?

23.2 Area of a Parallelogram

What is the area of a parallelogram with a height of 4 feet and a base of 8 feet?

23.3 Area of a Trapezoid

A trapezoid has a lower base of 12 feet, an upper base of 9 feet, and a height of 5 feet. What is its area?

23.4 Hypotenuse of a Right Triangle

What is the length of the hypotenuse of a right triangle with sides of 3 feet and 4 feet?

23.5 Law of Cosines for Triangles

Two sides of a triangle are 6 feet and 8 feet. The angle between the two sides is 34°. What is the length of the third side?

23.6a What is the area of a triangle with a base of 10 feet and a height of 12 feet?

23.6b What is the area of a triangle with two sides of 8 and 14 feet and an included angle of 23°?

23.6c What is the area of a triangle with sides that are 3, 4, and 6 feet?

23.7 Area of a Spherical Triangle

What is the area of a spherical triangle on a sphere of radius 100 feet with three angles that are 100°, 80°, and 70°?

23.8 Angles in a Polygon

What is the sum of the interior angles in a hexagon (6-sided polygon)?

23.9 Area of a Regular Polygon

What is the area of a regular polygon with 7 sides, each of which is of length 3 inches?

23.10 Area and Circumference of a Circle

What are the area and circumference of a circle with a radius of 4 feet?

23.11 Area and Arc Length of a Sector

What is the area of a sector of a circle with a radius of 3 feet with a central angle of 52°, and what is the arc length subtended by that angle?

23.12 Area and Angle Cut Off by a Chord

What is the area of the smaller portion of a circle with a radius of 4 feet that is cut off by a chord of 3 feet in length, and what is the central angle subtended by the chord?

23.13 Determining the Radius from a Chord

The length of a chord of a circle is 6 inches, and the distance from the center of the chord to the minor arc of the circle it determines is 2 inches. What is the radius of the circle?

23.14 Area and Circumference of an Ellipse

What are the area and circumference of an ellipse with a semimajor axis of 8 inches and a semiminor axis of 6 inches?

Chapter 24 Solid Geometry

24.1 Volume of a Solid with Constant Cross Section

What is the volume of a solid of constant cross section that has a base area of 12 square inches and a height of 16 inches?

24.2 Volume Remaining in a Horizontal Cylindrical Tank

A cylindrical tank that is 10 feet long and has a radius of 4 feet is lying on its side and contains liquid. The liquid level is 1 foot below the top of the tank. What is the volume of the liquid?

24.3 Surface Area of a Rectangular Box

A rectangular box is 8 inches long, 6 inches wide, and 12 inches long. What is its surface area?

24.4 Surface Area of a Cylinder

What are the lateral surface area and the total surface area of a cylinder that is 10 inches high and has a radius of 4 inches?

24.5 Volume of a Regular Pointed Solid

What is the volume of a regular pointed solid with a base that has an area of 50 square inches and a height of 12 inches?

24.6 Surface Area of a Pyramid

What are the slant height, lateral surface area, and total surface area of a 4-sided regular pyramid with 3-inch sides and a height of 10 inches?

24.7 Surface Area of a Cone

What are the slant height, lateral surface area, and total surface area of a cone with a radius of 6 inches and a height of 8 inches?

24.8 Volume and Surface Area of a Sphere

What are the volume and surface area of a sphere with a radius of 5 feet?

24.9 Volume of a Spherical Cap

What is the volume of a spherical cap 3 feet high from a sphere with a radius of 5 feet?

24.10 Volume and Surface Area of a Torus

A torus is formed by rotating a circle with a radius of 3 inches parallel to a central axis located 10 inches from the center of the circle. What are the volume and surface area of the torus?

Chapter 25 Conversions

25.1 Temperature Conversion

The temperature on a pleasant summer day is 25°C. What is the temperature on the Fahrenheit scale?

CHAPTER 27

 # More Than Just Formulas

Formulas Are Awesome!

There's no question that formulas are one of the greatest creations of mathematics. They represent the answers to myriads of questions, expressed in a single succinct (well, mostly succinct) expression. Whether it's Einstein's iconic $E = mc2$, which tells us how the stars produce the energy that supports life, or a much more humble formula such as total cost = number of items × cost per item, formulas not only give us insight and answers, they perform the equally important functions of making it no longer necessary for us to reinvent the wheel every time a particular problem needs solution.

Formulas are so awesome that almost every other discipline uses them. Other than language, there may be no single creation of man whose use is so widespread. And formulas are only a little harder to use than language. Every child learns a language (some in bilingual households learn two at once), and almost every child learns mathematical formulae only a few years later.

Almost everyone – especially students – comes to appreciate how powerful and useful formulas are. As a result, many acquire the attitude that all mathematics can be boiled down to formulas. In fact, the second-most common question of students is "What formula do I plug this into?" The most common question, of course, is "Will **this be on** the test?"

Beyond Formulas

Powerful though formulas are, they are easily forgotten (unless one is planning on a career that uses formulas extensively) once the test is past. In fact, even though algebra is a required subject for graduation in most high schools, the great majority of students who take algebra will never use a single algebraic formula in their life. They'll almost certainly never factor a polynomial, and if they have to solve an equation, it will generally be a relatively simple one, and they probably will do very little of that.

So why even bother to take algebra? If you're never going to use it, why bother with it.

The Power of Mathematical Reasoning

All mathematics courses use reasoning, and the ability to reason is one of the most useful tools that any individual can acquire. Those who defend algebra (which often finds itself under attack) point to the fact that algebra is a closed system in which the facts and assumptions are well defined, and even though the conclusions (such as the quadratic formula) and the techniques of algebra may be forgotten, the ability to reason in such a system is so important that it transcends the products of that reasoning.

Sometimes the reasoning itself is tremendously simple, but the idea underlying that reasoning is itself a powerful tool. And that's the case in the following story.

The Hole Truth

I first saw this story in one of the best math books I've ever read – How Not To Be Wrong, by Jordan Ellenberg. It is a book that is devoted not to formulas – although there are some – but to the importance and power of mathematical reasoning. The story below is the first anecdote in the book.

World War II saw the application of mathematical reasoning in a number of different areas, as the Allies sought to gain an advantage against the opposing powers. In England, a team of mathematicians was assembled, headed by the brilliant Alan Turing, in order to crack the German cryptographic machine Enigma. In the United States, mathematicians, physicists, and engineers gathered in Los Alamos for the Manhattan Project, which built the atomic bomb that eventually ended the War.

Meanwhile, back in Manhattan (New York's most famous borough), the Army had assembled the Statistical Research Group, which was housed near Columbia University. The SRG was charged with answering a number of diverse questions relating to the application of statistical analysis to the war effort. One of the problems that confronted the Army was: where should we put additional armor on our airplanes in order to minimize the chances that they would be downed by enemy gunfire? You couldn't put too much armor on a plane; it would never get off the ground. But where do you put the armor where it will do the most good?

The Army had compiled statistics on the bullet-hole density in various portions of those planes that returned from a mission (obviously, they couldn't get statistics on those planes that had been shot down). They showed the following table to the SRG

Section of Plane	Number of Bullet Holes per Square Foot
Engine	1.11
Fuselage	1.73
Fuel System	1.55
Rest of the Plane	1.80

The SRG had an odd mixture of people working in it; one of whom was Abraham Wald, a Jewish refugee mathematician from Austria. Wald would later go on to become one of the great statisticians of his day, but he should be remembered for his solution to this problem; he told the Army to put the armor on the engine.

Why? he was asked.

Wald's reasoning was simple. Instead of looking at the actual data, take it to the extreme. Suppose that some section of the plane came back with zero bullet holes per square foot in it. What conclusion would you reach?

It's obvious when you think about it. If every plane that came back had NO bullet holes in the engine, the obvious reason is that every time a single bullet hit the engine, the plane went down. Therefore, the engine was especially vulnerable. True, the data didn't show exactly that, but the reasoning showed that the engine was more vulnerable than any other section of the plane.

Another way of reaching the same conclusion can be found by supposing that all the other numbers were much higher. Instead of numbers between 1.55 and 1.80, suppose instead that they were all much bigger; possibly 10 or higher. The conclusion you would reach from looking at it this way is that the other sections of the plane were much more durable; like the Timex watches of old, they could take a licking and keep on ticking.

Possibly some readers of this book are descendants of pilots who owe their lives to the mathematical insight of Abraham Wald. This bit of mathematical reasoning didn't win the War – but it certainly helped.

It never hurts to look at extreme situations. Which brings us to a problem that may very well confront you today.

Should I Vaccinate My Children?

As I write this section of the book (in January, 2015), there is currently an outbreak of measles in California, which has been traced to a number of occurrences in people who visited Disneyland. Up until a few years ago, measles had roughly the same status as smallpox and polio; it had been effectively eradicated thanks to an intensive vaccination program. Then, in 1998, an article appeared in the prestigious journal Lancet, in which the MMR (measles, mumps, rubella) vaccine was erroneously connected with the appearance of behavioral disorders in 8 out of 12 children in a study group. I say "erroneously" because the article was later retracted.

Unfortunately, the damage had been done. A fairly well-known celebrity whose child had autistic symptoms and who had been vaccinated with MMR vaccine conducted a high-profile campaign against vaccination. As with the case with many celebrities with media access, her words were heard and unfortunately heeded by many who chose to pay attention to a celebrity rather than to medical authorities. There are sections of California currently in which more than 20% of the children have refused to be vaccinated.

So let's see what Abraham Wald would have said about this. First of all, he would have pointed to the fact that in the two extreme cases, when everyone was vaccinated, there were almost no cases of measles – and in the era before the vaccinations were available, measles were a common childhood disease. He would also have asked what's the worst that could happen, and how likely (that's what statistics is about) is the worst case to happen.

He would have discovered that, even under the erroneous assumption that there was some risk of autism with vaccination, there was a more significant risk from contracting measles later in life than childhood. Pneumonia and encephalitis occurring in conjunction with measles can result in death.

There are many examples where the reasoning exhibited in mathematical situations have applications far beyond what one originally sees in the classroom. And that's one of the reasons that educators feel that algebra is so important – as are other courses in mathematics. Important though formulas are, mathematics is more than just formulas. It's not just "What formula do I plug this into?" It's "What does this tell me that I can use in my daily life?"

And there's a lot – as the story of Abraham Wald and the bullet holes shows.

ABOUT THE AUTHOR

Jim Stein is a Professor of Mathematics at California State University in Long Beach. A graduate of Yale University with a doctorate from the University of California at Berkeley, Jim has written 30 technical articles on mathematics and several books explaining mathematics for general audiences.

His books *How Math Explains the World* and *How Math Can Save Your Life* were selections for the Scientific American Book Club. His book *The Right Decision* resulted in his being invited to be a guest blogger for the Huffington Post and Psychology Today. His next book, *Cosmic Numbers: The Numbers That Define the Universe*, has received critical praise and will be published by Basic Books, the leading publisher of scientific books for the general public.

Jim has also conducted conferences for the National Science Foundation and served on textbook adoption committees for the State of California. He has been an advisor on mathematical education at both the state and national level.

Jim also has experience related to mathematics that goes beyond the world of academics. He worked for an engineering firm on projects connected with putting a man on the Moon. He has also developed models for financial analysis and is the co-author of *How to Shoot from the Hip Without Getting Shot in the Foot*, a book on business management strategy.

INDEX

U

W